THE GREAT RECONFIGURATION

INTELLIGENCE RISING

The Platform Shift That Changes Everything

SRULY ROSENBERG

OOretz Press

The Great Reconfiguration series. Published by OOretz Press, an imprint of OOretz Inc.

"Sruly Rosenberg" is the pen name of Yisroel Rosenberg.

First edition, 2026. Revised text (v1.1), July 2026.

ISBN 979-8-951578-00-6 (paperback) · ISBN 979-8-951578-01-3 (hardcover)

The state of artificial intelligence described in this book reflects the field as of mid-2026. Capabilities, model names, and market conditions are moving quickly; treat specific figures as a snapshot, not a forecast.

For the ones who build before the road is paved —
and the ones who decide to take it.

Every great shift installs a tollbooth before it builds the road. The ones who pay early aren't lucky. They're early.

— S.R.

CONTENTS

INTRODUCTION

THE TOLL OR THE ROAD

IN THIS CHAPTER the one question underneath every decision you'll make for the next decade — *will you own a road or pay a toll?* — and the two-hundred-year pattern that decides the answer. **Why it matters:** the pattern has run six times and never once spared the incumbent who misread it. **What to do with it:** find out, before the chapter ends, whether you are building the rails or about to rent them.

There is a question underneath every decision you will make for the rest of your career, and almost no one in your position is asking it out loud.

It is not *"How do I use AI?"* That question is already beneath you. It is the one your competitors are answering, and answering it well will keep you precisely where you are while the ground moves underneath you. The real question is colder, and older than software:

When this shift is finished, will you own a road — or will you pay a toll?

Every platform shift for two hundred years has sorted the powerful into exactly those two groups. And it has never once cared who was bigger walking in.

Let me show you how fast *bigger* becomes *gone*.

The Call That Wasn't Real

It is a Tuesday in Hong Kong. A finance clerk at a multinational joins a video call with his chief financial officer and three senior colleagues — faces he has known for years, voices he would recognize in a dark room. They walk him through a series of urgent wire transfers. Routine, for a deal this size. He raises a small doubt; the email that set up the call had felt off. But he is looking at his CFO and his colleagues, live, answering his questions. So he sets the doubt down and executes. Across that day, fifteen transfers totaling **$25.6 million** leave the company's accounts.

Every person on that call was synthetic.

The CFO was a real-time deepfake. The colleagues were fabricated from public video, rendered well enough to fool a trained professional who had sat across from these people for a decade. He checked. He recognized. He asked questions and received fluent answers. None of it saved him, because the reality he was evaluating was not real. This was the engineering firm Arup, in early 2024. No one has been arrested. The money was never recovered.

Here is the part your strategy team will miss. The lesson is not *"AI fraud is rising."* The lesson is structural: **when reality itself can be manufactured, trust stops being a feeling and becomes infrastructure — something that has to be built, verified, metered, and owned.** And the iron law of every platform shift is simple enough to carve into a cornerstone: *whoever builds the infrastructure that everything else is forced to depend on captures the value of everything else.*

That is not an observation. It is a blueprint. And it has executed, without one exception, six times in the last century and a half.

You Are Not the Candlemaker

You already know the candlemaker loses. Every book about disruption casts you as the candlemaker so you'll feel the danger. I am going to do the opposite, because it is more honest about who you actually are.

You are not the candlemaker. You are the **gas-lighting executive** in 1880: established, capitalized, admired. You hold municipal contracts and thousands on the payroll. When a man named Edison lights four hundred bulbs along a half-mile of lower Manhattan, you run the math any competent operator would run. Four hundred bulbs against your hundreds of thousands of gas lamps. A rounding error. A parlor trick in one neighborhood.

You are exactly right about the snapshot, and dead wrong about the trajectory. And the trajectory is the only number history will ever record.

Within forty years your invincible industry is *gone*. Not because gas stopped working. Because someone built a better road, set the standard for it, and made the whole world pay to use it.

History remembers Thomas Edison as the man who invented the lightbulb. He did not. At least twenty-three inventors built working bulbs before he touched the problem. Joseph Swan had a working carbon-filament bulb on display by 1878 and lit his own house with it, though his key British patent did not come until 1880. Henry Woodward and Mathew Evans patented a bulb in 1874 and, unable to raise the money to build anything around it, sold their U.S. patent rights to Edison in 1879 for the commonly cited sum of five thousand dollars. They all solved the *invention*. They all lost the *infrastructure*.

What Edison did was the unglamorous thing none of them could do: he **finished.** He set the standard: the screw socket still in the ceiling above you, unchanged, 145 years later. He laid the rail. Pearl Street Station, the copper under the streets, the meters, the billing systems, the trained crews who explained the terrifying new technology to nervous customers. He did not sell light. He sold the *system that delivers light,* and then he owned the toll on every lamp that would ever plug into it.

Invent, and you earn a footnote. **Finish, and you own the century.**

That move is the most repeatable act in the history of capital: let the inventor invent, then build the infrastructure, set the standard, and own the toll. It has a shape. We are going to name that shape, because naming a thing is how you stop being its casualty and start being its author.

It is called the **Durable Sequence.**

The Durable Sequence

Every platform shift in two hundred years has run through the same five stages, in the same order, with the same kind of winner at the end. The clock changes. Sometimes the cycle takes forty years, sometimes five. But the sequence never does.

1. **Invention.** The core technology is proven, and it is crude, expensive, and unreliable. A battery and two carbon rods. A neural network on two gaming GPUs in a Toronto lab in 2012. The world mostly ignores it, and the world is mostly right to. For now.
2. **Infrastructure.** The finisher builds the plumbing that makes the capability universal and cheap. Trenches, copper, meters. Data centers the size of small cities, and the power plants to feed them. This stage is boring, capital-hungry, and decisive. It is where the toll booths get poured.
3. **Platformization.** The infrastructure becomes the default the rest of the economy is built on. Costs collapse by orders of magnitude. Standards lock in. The screw socket wins. This is the open window: the brief season when positions are still available.
4. **Proliferation.** The platform becomes invisible, an assumption. Nobody decides to "adopt electricity"; it is simply *there*. Holdouts don't get argued with. They get bypassed.
5. **Consolidation.** The market winnows. Dozens of entrants collapse into two or three infrastructure networks that will own the toll for a generation.

Watch it run. In the 1840s, Morse proved the telegraph; **Western Union** built the network and took ninety percent of the traffic. In the 1870s, Bell proved the telephone; **Theodore Vail** built AT&T into a monopoly that ran American communication for a century. In 1903, the Wright Brothers proved flight; **Boeing and Douglas** built the airports, the supply chains, and the industry. The inventor gets the statue. The finisher gets the toll.

Six times. Electricity, the telephone, the automobile, computing, the internet, mobile. Six times the inventor was celebrated and the finisher was paid. You are standing in the seventh.

Where You Are Standing Right Now

Here is the only sentence in this book you need to memorize: **we are between Stage 2 and Stage 3 of the seventh turn, and that is the most valuable place to be standing in a hundred years.**

The infrastructure is being poured right now, in public, at a scale that has no precedent. In October 2025, **NVIDIA**, which sells the engines this revolution runs on, became the first company ever to cross a five-trillion-dollar valuation. The five largest cloud builders are spending on the order of **$700 billion in 2026 alone** on AI infrastructure: close to double what they spent the year before, and more, each, than most countries spend on their militaries. They are restarting decommissioned nuclear reactors to find the power. Microsoft signed a twenty-year deal to bring **Three Mile Island** back online, renamed and targeted for 2028 (later accelerated toward 2027), for no reason other than to feed the machines.

And the cost of the thing itself is falling off a cliff in real time. In January 2025, a Chinese lab called **DeepSeek** revealed a frontier-class model whose widely cited **$5.6 million** training figure – the reported compute for its V3 base model, not the all-in cost of R1 – landed at a moment when American labs were spending billions per run. In a single day, NVIDIA lost **$589 billion** in market value, the largest one-day loss by a single company in the history of the stock market. That was not a crash. That was the sound of intelligence repricing toward zero. Meanwhile the demand side has gone vertical: By May 2026, ChatGPT alone had more than **900 million weekly users** and was still approaching a billion, an adoption curve unmatched at that scale.

Read those facts the way a finisher reads them. Trenches are being dug. Meters are being installed. Standards are being set *this quarter*. The platform is not finished, which is precisely why there is still room to own a piece of it. By the time it is obvious to everyone, the positions will be assigned. They are being assigned now.

The Three Types

In every shift, people sort themselves into three groups, and the split barely moves across two centuries.

Ninety percent are Dismissers. They protect what works because it has always worked. They wait for proof. They move only when forced, and by then it is over. Kodak invented the digital camera and buried it to protect film. Blockbuster was offered Netflix for fifty million dollars and laughed. Nokia sold forty percent of the world's phones and was irrelevant within roughly five years of the iPhone. None of them were stupid. All of them were skilled, confident, and rational. Their dismissal was reasonable, and it was fatal.

Nine percent are Dabblers, and they are the *most* dangerous category, because they feel like they're doing something. They run pilots. They appoint a Chief AI Officer. They commission the strategy deck. But they will not cannibalize the profitable old model, so AI never touches a real workflow. Nokia had smartphone prototypes years before Apple. Blockbuster launched an online service and killed it to protect late fees. Dabbling, during a platform shift, is just a slow, expensive way to dismiss.

One percent are Finishers. They see the pattern before it is obvious, commit before it is comfortable, and build infrastructure others come to depend on. Microsoft owned the software layer every PC needed. Google organized the web until its name became a verb. Amazon ended up owning the ground half the internet runs on. They did not ride the wave. They built it, set its standards, and captured nearly all of its value.

Ninety. Nine. One. The top one percent captures more than the other ninety-nine combined. This is not a curve. It is a cliff. And which side of it you land on is not decided by talent. It is decided by *when you move.*

> **Sort yourself, honestly.** If you have "pilots," ask one question: has a single one changed a revenue-generating workflow? If not, you are a Dismisser with a budget, and the gap between the two is smaller than you think.

The Clock Is the Whole Story

Now the part that should change your calendar.

Every turn of the Durable Sequence has run faster than the one before it, because each new platform inherits the infrastructure of all the ones before. Electricity took roughly **forty years** to go from invention to default. Computing took about **fifteen.** The internet, about **ten.** Mobile, about **five.** AI does not need to wire the streets; electricity did that. It does not need to lay the fiber; the internet did that. It propagates instantly through the billions of screens already in your customers' pockets. Its window is on the compressed end of that curve: **three to five years, total, from ignition to consolidation.**

The ignition was November 30, 2022: the morning the capability became visible to anyone, not just technologists. Run the arithmetic from there. We are roughly **three and a half years into the window as of mid-2026.** Even the *generous* assumption, that AI gets the full five years of the three-to-five-year clock it is tracking, half the internet's roughly ten, closes it around **2028.** The compressed assumption closes it sooner. From where you stand, both calculations converge on the same uncomfortable answer: **the decisive phase is the next twenty-four months.**

This is why every planning horizon you inherited is wrong. In the electricity era, an operator who began adapting in 1885 still had twenty years of runway. A business owner who pivoted to computing in 1985 had a decade. A company that reacted to the iPhone in 2009 had three or four years. In this shift, a three-year roadmap is not a plan. **It is a four-times-too-slow surrender, written in advance.**

The window does not close with an announcement. It closes the way platforms always harden: standards congeal, procurement documents start naming specific interfaces, regulators write rules that assume the winner already exists, and integration gets cheaper for the natives and more expensive for everyone else. Those tells are already visible at the edges. When they are visible everywhere, the toll booths are built and the positions are assigned.

The New Arithmetic

For two centuries, professional and corporate worth obeyed one formula:

Value = Time × Skill.

Seniority was hours served. The moat was the apprentice layer: the thousands of repetitions that turned a junior into an expert, and that no competitor could skip. The entire structure of firms, salaries, and careers was built on the scarcity of those repetitions.

The machine has just industrialized the repetitions. When intelligence itself becomes abundant, when expert-grade analysis, drafting, and reasoning cost fractions of a cent and arrive in seconds, the premium for raw execution collapses toward zero. Not slowly. The way the price of light collapsed once the grid was built.

So the formula changes:

Value = Intent × Leverage × Judgment.

You are no longer paid for the manual execution of tasks. You are paid for the precision of your **Intent:** knowing exactly what to build and why. For the scale of your **Leverage:** the systems and agents you command instead of the hours you log. And for the accountability of your **Judgment:** standing behind the output, signing it, being the real and answerable human at the edge of an automated loop. The machine performs. It does not get to be *responsible*. That is the one thing it cannot take, and therefore the one thing worth being.

The organizations that win the next thousand days will not use AI to run yesterday's processes faster. They will build what amounts to a company operating system, a thin ring of human judgment around an automated interior, and they will out-ship institutions ten times their size for the cost of a few subscriptions. The old org chart paid the doer. The new one rewards the builder of the rails.

Which returns us to the only question that was ever on the table.

When this shift is finished — and it will be finished inside the window you are standing in — the rails the economy runs on will either be yours, or someone else's that you rent for the rest of your working life.

The candlemaker's choice was never about candles. Yours was never about AI.

Turn the page. Let's make sure the rails are yours.

Key Takeaways

- Every platform shift for 200 years runs the same five-stage **Durable Sequence,** and the *finisher* who builds the infrastructure, not the inventor who proves the idea, captures nearly all the value.
- We are between Stage 2 (Infrastructure) and Stage 3 (Platformization) of the seventh turn: the single most valuable place to stand in a century, because the positions are still open.
- People sort 90 / 9 / 1 into Dismissers, Dabblers, and Finishers. Dabblers don't lose to AI; they lose to Finishers.
- The window is the shortest in history: **~3–5 years, opened November 2022, decisive phase through ~2028.** A three-year roadmap is a four-times-too-slow surrender.
- The new arithmetic is **Value = Intent × Leverage × Judgment.** You are paid for what you specify, what you command, and what you can be held accountable for, not for execution.

Finisher Move: Stop comparing AI to what you do today. Start asking where, in your industry, you could own a piece of infrastructure everyone else will be forced to route through.

Sources & Currency Note

This book states facts as of mid-2026; specific figures move, the pattern does not. Key data points in this chapter: the Arup deepfake fraud (~$25.6M, reported early 2024, CNN/Hong Kong Police); NVIDIA as the first company to cross a $5 trillion valuation (October 2025); the ~$589 billion single-day NVIDIA loss following DeepSeek's R1 disclosure (January 27, 2025 — largest one-day loss by a single company in market history; DeepSeek's reported ~$5.6M training cost (its V3 base-model compute, not R1's all-in cost)); 2026 hyperscaler AI capital expenditure on the order of ~$700 billion (Big-5 cloud providers); the Microsoft–Constellation agreement to restart Three Mile Island Unit 1 (announced Sept 2024, restart targeted for 2028, later accelerated toward 2027); and ChatGPT weekly active users (more than 900M by May 2026 and still approaching ~1B, OpenAI). Where a single figure could age badly, it is framed as a range or dated explicitly.

CHAPTER 1

THE DURABLE SEQUENCE

IN THIS CHAPTER the five-stage law every platform shift has obeyed for two hundred years — *Invention → Infrastructure → Platformization → Proliferation → Consolidation* — proven across five industries that never once broke the order. **Why it matters:** AI is running the same sequence faster than anything before it, and the stage you are standing in tells you exactly how much time you have left to claim a position. **What to do with it:** stop asking whether AI is real and start locating your industry on the sequence — because the gap between two stages is where every fortune in this book was made.

THE DURABLE SEQUENCE

Five stages. Two hundred years. One pattern.

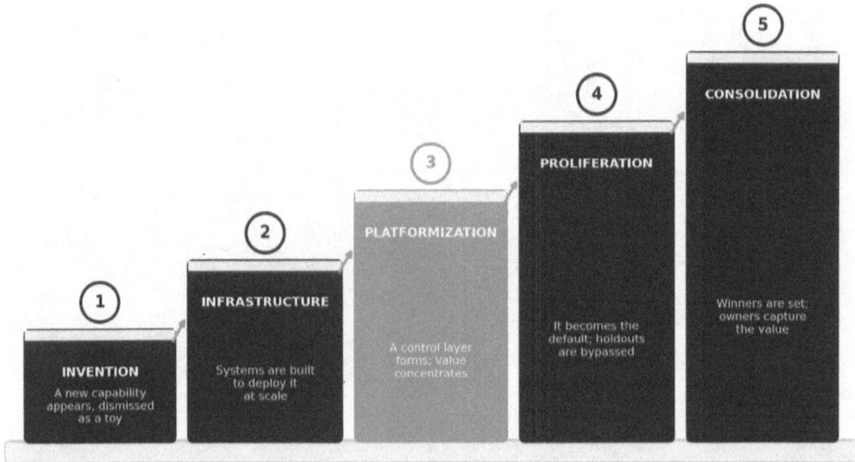

The inventor proves it. The finisher builds the layer that captures it.

It is 1876, and you are the best telegraph operator in New York City.

Your fingers move across the brass key at forty words a minute, faster than anyone in the office. Eight years went into that speed: learning to hear the difference between dots and dashes by their rhythm alone, to decode an incoming message while your hand sends a different one out. You earn three times what a factory man earns. Firms compete for operators with your accuracy. By any measure that exists, you are a master of your craft.

One afternoon your supervisor mentions something he read in the paper. A man named Alexander Graham Bell has shown a device that carries the human voice down a wire. You can talk to someone miles away as if he stood beside you.

You laugh, and your laugh is correct. You have heard about these demonstrations, parlor tricks for investors. The sound is dreadful; you can barely pull words out of the static. And a trained operator moves structured information faster and cleaner than anyone fumbling through a conversation ever could. Morse is precise. Voice is sloppy. Your skill is irreplaceable.

Within fifteen years your industry is obsolete. Your eight years of training are worth nothing. Your forty words a minute mean nothing. The telephone does everything the telegraph does, plus things the telegraph cannot imagine, and it asks for not one hour of specialized training. Anyone who can speak, literally anyone, can use it.

You could not see that in 1876, because in 1876 the telephone was worse. More expensive, less reliable, lower capacity. By every rational measure available to you, your dismissal was the sound judgment of a sound professional.

That is the first thing to understand about platform shifts, and the reason they keep killing competent people: **they do not arrive as a threat. They arrive as a bad bet you are right to refuse, until the day the bet was the only one that mattered.** The operator did not wake to find his job gone. He woke on a thousand mornings, each slightly worse than the last. Fewer messages. A little less urgency in the room. The hire they didn't replace. The branch that closed. The raise that didn't come. The slow certainty that the world had relocated to somewhere his skill set could not follow, because his skill set was engineered for a world that was quietly ceasing to exist.

Western Union still stood in 1900. Operators still had jobs. But the trajectory was already set in iron: telephone traffic outran telegraph traffic by an order of magnitude, and by 1910 it was not close. By 1960 the telegraph was a museum piece, a technology that had wired a nation, built fortunes, and employed tens of thousands of skilled professionals, reduced to a curiosity.

The operators who read the pattern in 1880 and moved, into telephone installation, switchboards, electrical engineering, did fine. The ones who clung to the brass key were the last generation of their profession.

This is why I open with the man who lost rather than the man who won. **You are not deciding whether AI is real. You are deciding, right now, which generation of your profession you intend to be: the one that read the pattern, or the one that was the pattern's final casualty.** Everything in this chapter is built to make that decision impossible to get wrong.

The Five Stages, As Law

In the Introduction I named the **Durable Sequence** and gave it to you in outline. I am not going to re-teach it. I am going to prove it. I will walk it through five industries and two centuries until you can recognize it operating on you in real time, because a framework you can only recite is decoration, and a framework you can see running is a weapon.

The Sequence has five stages. They occur in this order, every time, in every major platform shift in modern history. Not usually. **Every time.**

1. **Invention.** Several people prove the technology works. It is crude, costly, unreliable. Nobody gets rich. Inventors collect patents and citations. The market is tiny and fragmented, and the smart money stays away, correctly, for now.
2. **Infrastructure.** A finisher builds the systems that make the capability usable at scale: power plants, distribution, manufacturing lines, cloud platforms, APIs. This is where the real money begins to move, and it moves

overwhelmingly to the finisher, not the inventor. This is where the toll booths get poured.

3. **Platformization.** Costs collapse by orders of magnitude. Adoption detonates. An ecosystem of companies builds on top of the infrastructure. Network effects engage: more users make the platform more valuable, which pulls in more users. Standards lock. The window is open here, and only here.
4. **Proliferation.** The technology becomes the default assumption. New businesses are built presuming it exists. Holdouts are not argued with; they are bypassed. The old world does not fade. It falls off a cliff. One year the industry debates whether the thing matters; three years later everyone who didn't move is insolvent or irrelevant.
5. **Consolidation.** The winners are obvious. The platform owners have captured enormous value. Standards harden into permanence. The cycle completes, until the next shift begins it again.

Memorize the shape, not the dates. The clock changes (the cycle ran forty years for electricity and is running three to five for AI) but the order has never once rearranged itself. What follows is the proof, in five industries that never broke it.

Electricity: The Template for Everything After

We met the electricity story in the Introduction: Edison, the twenty-three prior inventors, Pearl Street. I am returning to it because it is not merely an example of the Durable Sequence. It is the master copy. Every shift after it is a faster pressing of this same disc.

Invention ran almost eighty years, from the early 1800s to 1879. Volta, Ampère, Faraday laid the science. The telegraph proved electricity could pay. More than twenty inventors built working bulbs. None of them could convert invention into infrastructure. They proved electricity worked. Not one of them made it inevitable. Hold that distinction. It is the entire chapter in five words.

Then in 1879 Edison opened the **Infrastructure** stage, and he did not do it by building a better bulb. He built the system underneath the bulb: power plants, distribution lines, meters, billing, the sockets, the switches, the fuses. **Pearl Street Station went live in 1882** and delivered current to lower Manhattan. The platform was born, not in a laboratory but in the ground under a city.

And here the story hands you a lesson most retellings skip, one aimed straight at anyone building infrastructure today.

Edison bet on the wrong standard. He built on direct current, which bled away over distance and could not be carried much past a mile, so every neighborhood needed its own plant. It worked. It could not scale. His rivals, George Westinghouse and Nikola Tesla, bet on alternating current, which could be stepped up for long-distance transmission and stepped back down at the home. One plant could light an entire city.

Edison fought AC with everything he had: propaganda, public electrocutions of animals, a campaign to wire the electric chair with AC so the public would couple it with death. None of it held, because the infrastructure economics were decisive and economics does not care about marketing. DC needed a plant every mile; AC needed one every hundred. City after city chose AC. In the end Edison's own company switched to AC after he lost control of it.

Read that as instruction, not trivia. **Being the finisher is necessary and not sufficient. You also have to win the standard, and the standard is won on economics, not on who got there first or fought hardest.** Edison was the greatest infrastructure builder of his age and he still lost the standards war, because he anchored to the architecture he had already built instead of the one the math demanded. The AI standards war is being fought in this exact way right now: which model interfaces become universal, which deployment patterns become assumed, which protocols everything else is forced to speak. The firms that set the economically inevitable standard will own the toll. The brilliant ones that anchor to yesterday's architecture will lose with their genius fully intact.

Platformization struck between roughly 1895 and 1910, and it struck like a hammer. Cost per kilowatt-hour fell from Edison's twenty-four cents to around a dime – a collapse of more than half in nominal terms, far steeper in real terms – and electrification of urban buildings went from a rounding error to roughly a quarter of them. The ecosystem erupted: once a home had current, entrepreneurs raced to build things that drank it. Fans, irons, stoves, sewing machines, doorbells, vacuum cleaners, air conditioning, washing machines, each one quietly rebuilding daily life.

Then came the moment that turned electricity from interesting to inevitable. In 1893 Westinghouse won the contract to light the Chicago World's Fair, and the scale was staggering: roughly a hundred thousand incandescent lamps, more electric light in one place than had ever existed on Earth. Some twenty-seven million people came through, most of whom had never seen an electric light in their lives. They walked into the Court of Honor at night and saw a city built of light: buildings outlined in white fire, fountains glowing, darkness shoved back as if by sorcery. These were people whose evenings were shaped by the flicker of flame, who slept when it went dark. And here was a world without darkness. They went home to their gas lamps and kerosene, and they wanted what they had seen. Councils that had been debating whether to invest stopped debating, because their citizens had now seen the future with their own eyes and wanted it in their own streets.

The Fair did not prove electricity worked. Edison had done that a decade earlier. **It proved electricity was desirable, that once people felt it they would never go back, and desire is what makes a platform unstoppable.** Every platform shift has a World's Fair moment: the demonstration that carries the technology out of the expert class and into the public imagination. For AI, that moment was

ChatGPT's launch on November 30, 2022, an adoption surge without precedent at the time. Like the fairgoers of 1893, people saw what was possible, went home, and wanted more. They are still wanting more; **by May 2026, ChatGPT had more than nine hundred million weekly users and was still approaching a billion.**

Network effects then closed the loop with brutal patience. More homes with current meant more customers for appliances, which drove appliance prices down, which gave more people reason to electrify. More customers per utility meant lower cost per kilowatt-hour, which made power cheaper, which pulled in more customers. By 1910 the momentum could not be stopped by anyone.

Proliferation, from about 1910 to 1925, killed the old world outright. The gas-lighting business that looked invincible in 1880 was gone by 1920. In 1890 the companies were still serene: electricity is a fad for the rich. By 1900 they were desperate and cutting prices. By 1910 most had quit the fight, pivoting to heating and cooking or merging into the very electric utilities that beat them. By 1920 gas lighting was a footnote.

But the deeper transformation was in the factory, and this is the parallel you cannot afford to misread. Before electricity, a factory ran on one large steam engine connected to every machine through a maze of line shafts, pulleys, and belts. The engine sat at the center, usually the basement, and the whole plant was arranged around it. Break a belt and the factory stopped. Add a machine and you had to solve how to chain it to the shaft. **The building was designed around the power source, not around the work.**

When electric motors arrived, each machine got its own. Now the plant could be designed around the work: machines placed where the workflow wanted them, started and stopped independently, the floor reorganized from a rigid central system into a flexible distributed one. And here is the fact that should reorganize your strategy: **the gains did not come for nearly two decades.** Factories that simply bolted a motor where the steam engine had been saw almost nothing. The leap, on the order of a several-fold jump in productivity, came only when plants were redesigned around what electricity made possible. Not faster machines. A new architecture of work.

The parallel to AI is not loose. It is exact. Today's organizations are arranged around a centralized source of intelligence, human experts, and everything flows through them, the way every belt once flowed through the steam engine. **Bolt AI onto that unchanged layout and you will get the same near-nothing those early factories got. That is the cost wall, and we will dismantle it in Chapter 3.** The several-fold gain arrives only for the organizations that redesign the work around distributed intelligence. The productivity does not live in faster thinking. It lives in the new architecture.

By the **Consolidation** stage, around 1925, the winners were named: General Electric, Westinghouse, the major utilities, all sitting on enormous captured value. The losers, gas companies, candlemakers, kerosene men, were gone. New work had replaced old: electricians, plant operators, appliance manufacturers, electrical engineers. The platform was mature. The standard was permanent. The cycle was complete.

Forty years, invention to default. Write that number down. You are going to watch it collapse.

The Telephone: When the Network *Is* the Product

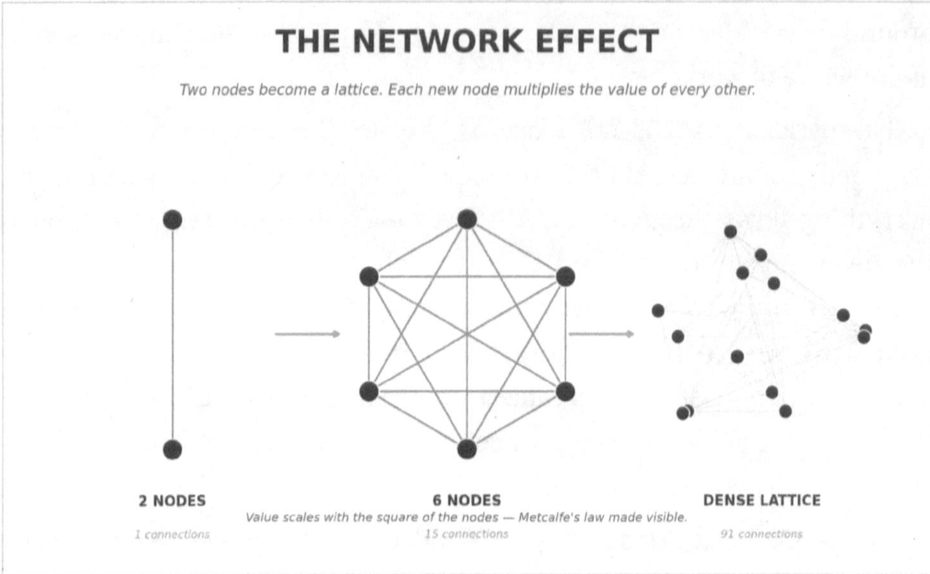

The operator who laughed at Bell made the precise error every incumbent makes in a shift: **he judged the new technology on the old technology's scoreboard.**

On the telegraph's terms, the telephone was inferior, and the operator's metrics were honest ones. The telegraph was faster for structured data, more reliable, lighter on bandwidth; a skilled hand cleared more messages an hour than any clumsy call. But the telephone was not playing on those terms. It was writing new ones. No Morse to learn. No skilled operator at either end. No wait for transcription. A real conversation: tone, emotion, interruption, laughter. **The telephone did not transmit information faster. It transmitted humanity, and there was no column on the telegraph's scoreboard for that.**

The **Invention** stage was a photo finish. Bell and Elisha Gray filed for similar patents on the very same day, February 14, 1876, and Bell reached the office hours ahead. Had Gray arrived first, every AT&T phone would have been a Gray phone. The history of technology turns on margins this absurd. But Bell won the legal battle and still did not win the platform, because winning the patent and winning the platform are different victories. Bell understood the technology and never understood the business.

Infrastructure belonged to **Theodore Vail**, who ran AT&T and grasped what Bell never did: telephony was a natural monopoly. Every new connection made the network more valuable to everyone already on it. A telephone is useless if you are the only owner, mildly useful at a hundred owners, transformative when everyone has one. This was one of the first clean demonstrations of the network effect, the same force that would later hand dominance to Facebook, Google, and Amazon, and the same force compounding under AI right now: more businesses adopt, more tools get built, the tools get cheaper and better, which pulls in more businesses, and the flywheel turns until adoption is no longer a decision.

Vail's strategy was simple and ruthless: **connect everyone, even where the short-term math says don't.** String copper to Kansas farms and Maine fishing villages where the revenue might never cover the wire, because once everyone is connected, the network becomes so valuable no competitor can replicate it. He was right. AT&T ran American communications for more than a century and was worth on the order of a quarter-trillion dollars at its peak.

And then the platform did what dominant platforms eventually do: it triggered a political response. The Bell System was not beaten by a better telephone. It was dismantled. The government's antitrust case ended in the divestiture that took effect in 1984, with AT&T keeping long-distance and a shrunken core while the regional "Baby Bells" inherited the networks.

The brand on the door hides a twist. The AT&T on signage today is not the original. One of the Baby Bells, SBC, bought the weakened parent in 2005 and put on its name. **The child bought the parent and wore its face.** So Vail's logic about infrastructure dominance was sound, and the more brutal lesson sits right beside it: **even the greatest finisher can be broken up, reassembled, and renamed.** "AT&T survived" is true as a logo and false as an unbroken company. Bookmark that. By the end of this chapter it becomes the rule that no position, however total, is permanent.

In the telephone's first era, **Platformization** was already remaking the world, and it did it the way every real platform does: not by improving what existed but by creating what could not have existed. Before the telephone, "managing a business" meant coordinating only with people you could visit or write to. A Chicago factory owner checking on a Pittsburgh supplier could travel or mail a letter and wait days. The telephone cut that to minutes, and the modern multi-location corporation, the organizational form that dominates the global economy today, became possible for the first time in human history. Not existing firms getting more efficient. **Entirely new structures coming into being because instantaneous voice made them feasible.** The same in private life: a daughter who moved away was once a letter a month; now she was a voice on demand. Courtship changed. Friendship changed. The telephone did not transmit information across distance. It dissolved distance as a barrier to human connection.

That is the lens for AI, and it is the one almost everyone gets wrong. **AI is not improving existing workflows by ten percent. It is creating categories of work, creativity, and problem-solving that were not possible before any human had an on-demand reasoning partner available at all times**, the way the multi-location corporation was not possible before the call. If your AI strategy is a list of tasks to do faster, you have mistaken the telephone for a quicker telegraph.

And Western Union, the telegraph monopoly that looked unkillable? It had the chance to buy Bell's patents. It declined: the technology was inferior, the customers didn't want it, the market was too small. By 1900 the telegraph was in irreversible decline; by 1960 it was effectively dead. **The most rational analysis in the room was the fatal one.** Remember that the next time your own analysis feels airtight.

Aviation: When the Inventors Lose and the Industrialists Win

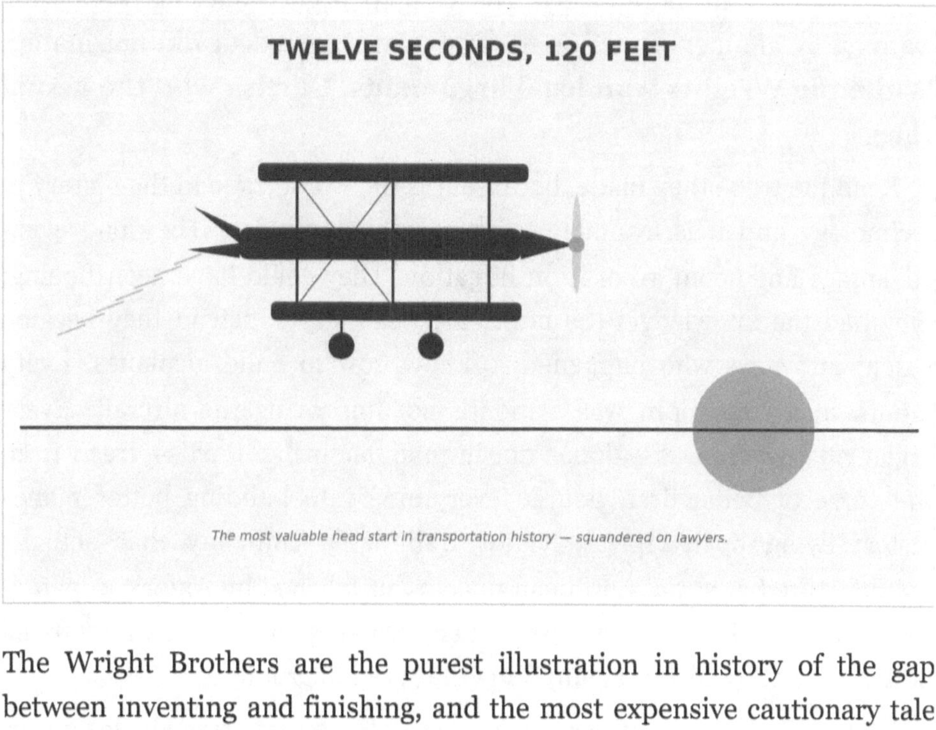

The most valuable head start in transportation history — squandered on lawyers.

The Wright Brothers are the purest illustration in history of the gap between inventing and finishing, and the most expensive cautionary tale ever told about confusing the two.

On December 17, 1903, at Kitty Hawk, Orville and Wilbur Wright achieved the first powered flight: twelve seconds, a hundred and twenty feet. A triumph of engineering, nerve, and persistence. They had solved one of humanity's oldest dreams and made a heavier-than-air machine fly. They were brilliant. They were first. They held the patents, the knowledge, the public's admiration.

Their company failed.

They spent years suing competitors who iterated on their designs, so consumed with protecting the invention that they never built the infrastructure that would make aviation practical. Their patents covered "wing warping," twisting the whole wing to steer, and they sued anyone who came near it, including Glenn Curtiss, who had built a better system: ailerons, small movable flaps on the trailing edge that did the same job more reliably and more safely. The Wrights argued ailerons were wing warping by another name. Courts sometimes agreed. It did not matter. **While the Wrights won legal arguments, Curtiss won the actual race.**

Read the trade they made, because it is the worst trade in the history of technology and it is available to you every single day. They had every advantage and spent all of it on litigation. They could have been Boeing; they had the knowledge, the fame, the head start. Instead they became patent attorneys who happened to know how to build airplanes. Every month in a courtroom was a month not improving the aircraft. Every dollar on lawyers was a dollar not in manufacturing. Curtiss, freed from the curse of being first, poured everything into building better planes faster: flying boats, naval aviator training, a company that actually manufactured at scale. European makers, untouched by American patent fights, advanced even faster; the French, British, and Germans were flying better aircraft by 1910 than anything the Wrights produced.

Wilbur died in 1912 at forty-five, worn out and embittered by legal war. Orville sold the company in 1915. The Wright Company was absorbed into larger manufacturers and vanished. The men who gave humanity flight left no lasting business, only a name in a museum. **They held the most valuable head start in the history of transportation and traded it for a stack of legal briefs.**

The companies that won aviation (Boeing, Douglas, Lockheed) understood that aviation was never about building planes. It was about building the system that made planes useful: manufacturing that produced reliable aircraft at scale, airports with runways and terminals and fuel, air traffic control that could keep dozens of planes safely in one sky, navigation, weather forecasting, pilot training, maintenance standards, regulatory frameworks. Boeing alone became a company worth hundreds of billions; commercial aviation generates well over a trillion dollars a year. The Wright Company closed its doors. Their names live in the history books and their business legacy is zero.

This is the pattern that should haunt every AI researcher publishing a paper today. **The people proving AI is possible will win academic immortality. The people building AI infrastructure will win the fortunes.** Both are honorable. Only one owns the toll. The question on the table for everyone in this field is not *can you invent.* It is *which one are you building toward.*

The Automobile: Ford's Triumph and Ford's Trap

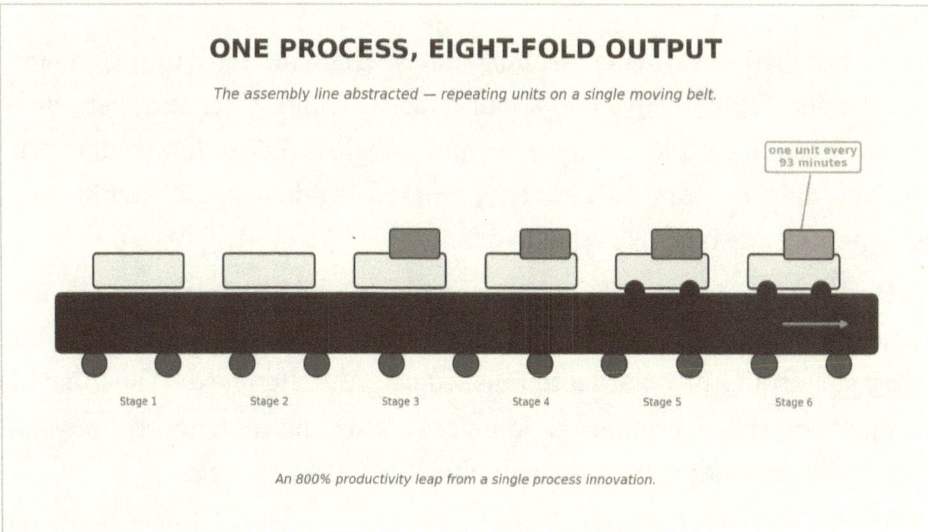

Before Henry Ford, automobiles were craft objects built by artisans. In 1908 there were more than two hundred and fifty American car manufacturers, together turning out roughly sixty-five thousand vehicles a year, at prices that, in today's money, ran well into the tens of thousands of dollars apiece. Under one percent of American families could afford one. The car was a toy for the ultra-wealthy.

Ford invented none of it. Dozens had built automobiles. He did not even invent the assembly line; Ransom Olds used one in 1901. **What Ford did was finish the system,** and he did it by removing options, not adding them.

He radically simplified the product. While competitors offered dozens of choices, Ford offered one model in one color. The Model T carried roughly fifteen hundred parts where the typical car had several thousand. Every component standardized. No customization, no options. It sounds like a limitation; it was strategic brilliance. By killing variation, Ford reached economies of scale no rival could touch. **File that, because it is the spine of this whole book: the finisher wins by subtraction, by making one thing universal and cheap, not by out-featuring the field.**

He installed the moving assembly line at Highland Park in 1913. Before it, a Model T took twelve and a half hours to build. After it, ninety-three minutes, an eightfold collapse from a single process innovation. The workers did not work harder. They worked inside a better architecture, each performing one task repeatedly.

He integrated vertically on a scale no one had attempted, with iron mines in Minnesota, rubber in Brazil, glass, steel, railroads, controlling every step from raw material to finished car. And then he did the thing his competitors called insane: on January 5, 1914, he announced he would pay five dollars a day, more than double the prevailing wage.

The Wall Street Journal called it "an economic crime." Rivals predicted bankruptcy. Nobody paid unskilled line workers that kind of money. Ford's logic was simple and devastating: **his workers were also his customers.** At the old wage a Ford man could not afford a Ford. At five dollars a day he could. By paying workers enough to buy the product, Ford manufactured the demand that fed his volume, which drove down cost, which made the car cheaper, which made more demand, a flywheel his cost-minimizing competitors could not even see. The higher wage also crushed turnover from a ruinous figure (high enough that he was hiring several workers for every position just to keep the line staffed) down to a fraction of it. A stable, motivated workforce more than paid for the raise. **It was a platform move disguised as a labor policy**, and noticing the platform move hiding inside an apparently unrelated decision is a skill this book will train into you.

The results were staggering. The Model T fell from over eight hundred dollars in 1908, when Ford built ten thousand of them, to under three hundred by the mid-1920s, when production passed two million a year. By the time the line stopped in 1927, Ford had sold more than fifteen million. American car ownership went from under one percent of families in 1908 to well over half by 1927, luxury to necessity in less than a generation. Ford had finished the automobile platform.

And the ripple effects dwarfed the car itself. The automobile did not just replace the horse. It created suburbs, because workers could now live miles from work; it created the highway system, the largest public works project in human history; it created the oil economy that rewired global geopolitics; it created consumer credit, because cars cost enough that people needed installment plans, which birthed the entire consumer-finance industry; it created fast food, shopping malls, drive-ins, teen culture. **It created industries worth more than the automobile industry itself. Exactly as AI will create industries worth more than the AI industry itself.** When you size the AI opportunity, you are almost certainly sizing the car and missing the suburbs.

And then the story delivers its most brutal lesson. **Ford fell into the Success Trap**, the failure pattern we anatomize in Chapter 3, where the very methods that made you win calcify into the prison that ends you.

By 1921 Ford held the majority of the American car market and was the undisputed winner. And he went rigid. No color but black. No financing. No annual model changes. No variety. "The Model T is perfect," he said. "Why would we change it?" Meanwhile Alfred Sloan at General Motors innovated on everything Ford refused to touch: marketing, financing, customer psychology, model changes, multiple price points, colors, options, offering "a car for every purse and purpose." The result was devastating: Ford's commanding majority of 1921 bled away year after year while GM's share surged past forty percent. Ford had to shut production for six months in 1927 to retool from the Model T to the Model A – the only stretch his share ever fell below ten percent – and though the Model A was America's best-selling car in 1929 and 1930, Chevrolet retook the lead in 1931 and Ford never saw his old majority again. **The finisher had been finished, by his own success.**

This is the most frightening lesson in the entire Sequence, and the reason no one reading this book gets to feel safe. **Winning the platform shift does not guarantee survival.** Ford's dominance felt permanent, a commanding market share, fifteen million cars, the most famous businessman in America, and in a handful of years it was gone, not because the industry declined (it was booming) but because Ford mistook his method for the product. The method was mass production. The product was what customers actually wanted. When those two diverged, his method became his prison. The same rigidity that destroys incumbents during a shift destroys the finishers after it. The Sequence does not stop because you have won. It simply starts looking for its next casualty, and a recent winner is its favorite kind.

Computing: IBM's Two Fatal Decisions

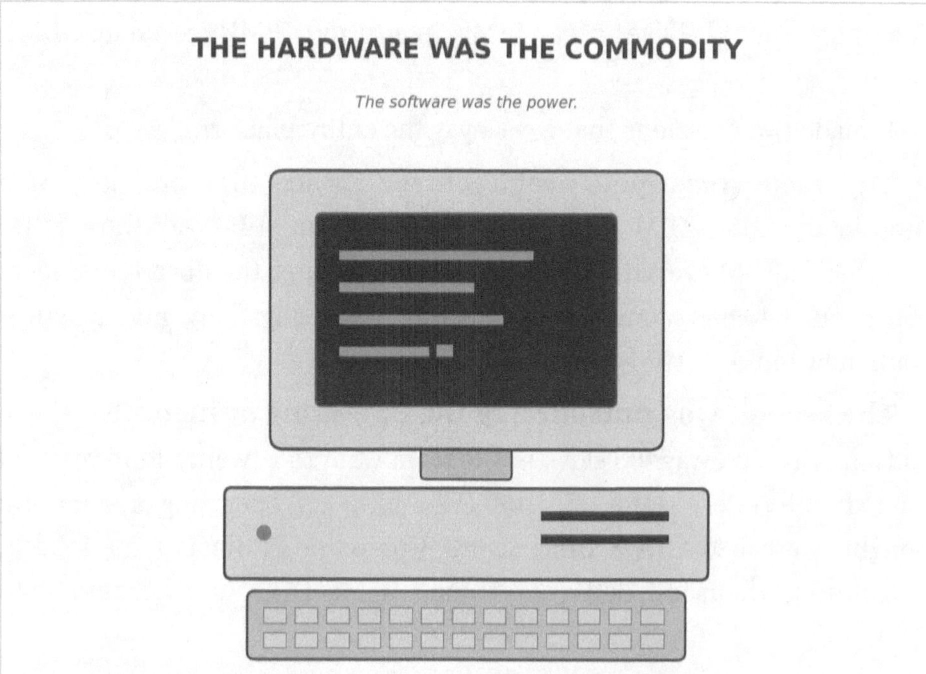

IBM built the PC. The platform — the operating system — captured the power.

In 1980, IBM was the most powerful technology company on Earth and arguably the most powerful company of any kind. It dominated mainframe computing with roughly seventy percent of the market. Its annual revenue ran north of twenty-five billion dollars, more than the GDP of many countries. "Nobody ever got fired for buying IBM" was not marketing; it was the operating reality of corporate America. IBM machines ran the banks, the airlines, the government agencies, the hospitals. If the mainframes went down, civilization hiccupped. The company was so dominant the U.S. government spent more than a decade trying to break it up.

When the personal computer revolution began, IBM held every advantage, the kind that makes competition look pointless. The engineering talent. The manufacturing capacity. The brand every corporate buyer trusted on reflex. The distribution. The relationships. If any company should have owned the PC era, it was IBM. The personal computer should have been IBM's platform, IBM's revenue, IBM's century.

It made two decisions that gave away the entire platform.

The first was open architecture. Earlier IBM products were proprietary: buy IBM hardware and you ran IBM software, IBM peripherals, IBM everything. But to ship the PC fast, the Boca Raton team chose off-the-shelf components anyone could buy. Any manufacturer could now build an IBM-compatible machine.

The second was outsourcing the operating system. IBM's own software division was too slow, so the team went to a twenty-four-year-old named Bill Gates. Gates did not even have an operating system. He bought one called QDOS from Seattle Computer Products for a total of seventy-five thousand dollars, renamed it MS-DOS, and licensed it to IBM.

Here is the detail the whole century turns on: **Gates *licensed* MS-DOS to IBM. He did not sell it.** He kept the right to license the same operating system to every other PC maker alive. IBM agreed because IBM did not think software mattered. The hardware was the product, the operating system a commodity component.

It was the single most consequential business mistake of the twentieth century, and it happened because IBM believed it was in the hardware business. It was not. **It was in the platform business, and it handed the platform away** for the price of a component it didn't respect.

Within a decade, dozens of companies were building IBM-compatible PCs (Compaq, Dell, HP, Gateway) all on the open architecture IBM created. Hardware became a commodity. Margins collapsed. Every manufacturer could build a PC, so no manufacturer had pricing power; the race to the bottom was swift and merciless. **But every one of those machines, every Compaq, every Dell, every no-name clone from**

a Shenzhen garage, needed MS-DOS, and later Windows. Gates owned the one piece of the system that could not be replicated, because he owned the standard. No factories. No distribution. No hardware. Just the layer everything else was forced to depend on. **Microsoft captured the platform while IBM gave away the commodity**, and by early 1993 Microsoft's market value had passed IBM's, the company that bought a seventy-five-thousand-dollar operating system now worth more than the company that built the personal computer. IBM sold its PC business to Lenovo in 2005 and left the market it invented, unable to win the commodity race it had accidentally created.

Now line it up. Twenty-three inventors made lightbulbs before Edison. Hundreds of companies made cars before Ford. The Wright Brothers flew before Boeing existed. IBM built the personal computer before Microsoft captured it. **The pattern does not rhyme. It repeats with the precision of a law.**

And computing adds the lesson that should reorganize how you look at your own industry tonight: **the platform is usually not the visible part of the system.** IBM's hardware was what people saw, touched, and bought. Microsoft's software was invisible, code running underneath everything. The invisible layer was the platform. The visible layer was the commodity. In AI the same dynamic is live: the models (ChatGPT, Claude, Gemini) are what people see, but the infrastructure beneath them (the GPUs, the data centers, the APIs, the cloud platforms) is where the platform power sits. We had a preview when crypto mining turned GPUs into industrial assets years before AI did, forcing the same questions AI now forces: where does the hardware come from, who controls supply, what happens when power and cooling become strategy, and what does it mean when a new workload turns a component into a platform. AI is far bigger than crypto. But crypto was the warning shot: when work becomes standardized and the payoff becomes real, the hardware becomes the battlefield.

> **PLATFORM RULE** In every shift, the invisible layer beats the visible layer. The thing customers see and buy is the commodity. The thing it is forced to run on is the platform. In AI, the infrastructure beneath the chatbots is where the power and the toll live.

The Pattern Applied: Where AI Sits Right Now

Five industries. Five eras. One Sequence, never broken. Now I am going to place AI on it with precision, because vague urgency is worthless and a coordinate is worth everything.

Every one of those shifts ran the same five stages in the same order. Every one saw inventors fail and finishers win. Every one destroyed incumbents who held every advantage. Every one compressed jobs while creating new ones. And every one ran faster than the shift before it. Here is AI's coordinate as of mid-2026:

Invention is complete. AI works. Large language models write, reason, code, and analyze at expert level. Computer vision works. Speech recognition works. Multimodal AI works. The hard problems were solved by dozens of brilliant researchers across hundreds of labs. They are this era's Davy, de la Rue, and Swan: the inventors who proved the concept, whose papers will be cited for decades and who will, if history holds, mostly not be the ones who get paid.

Infrastructure is in full swing. NVIDIA is building the engines, and **in October 2025 it became the first company ever to cross a five-trillion-dollar valuation.** Amazon, Microsoft, and Google are building the cloud. OpenAI, Anthropic, and Google are building the model layer. Hundreds of companies are building the tooling, the APIs and frameworks that let ordinary developers ship AI. **The big cloud builders are pouring capital in on the order of seven hundred billion dollars in 2026 alone, roughly double the prior year**, and they are restarting decommissioned nuclear plants to find the power. Microsoft signed a twenty-year deal to bring **Three Mile Island** back online, renamed and now targeted for 2027, for no reason but to feed the machines. **This is the Pearl Street moment.** The toll booths are being poured in public, at a scale with no precedent.

Platformization is just beginning. Costs are collapsing the way they always do at this stage. The price of frontier-grade AI has fallen by **more than ninety percent in roughly three years.** A top model that cost on the order of tens of dollars per million tokens at its 2023 debut now runs a few dollars or less for equal-or-better quality, with the cheapest capable models far below that, and it is still falling. **This is the electricity moment**, dollars per kilowatt-hour to single-digit cents. **This is the Model T moment**, eight hundred dollars to under three hundred. When cost collapses at this rate, adoption does not grow; it detonates. **By May 2026, ChatGPT alone had more than nine hundred million weekly users and was still approaching a billion**, an adoption curve unmatched at that scale, and the ecosystem is erupting: thousands of AI-native startups building on the platforms, exactly as thousands of appliance makers once built on the grid. Network effects are engaging: more users generate more data, which improves models, which attracts more users. The flywheel is turning. The full scale of the infrastructure race underneath these applications is its own subject, and we examine it in Chapter 6.

We are standing precisely at the inflection: the point where electricity went from Pearl Street to the Chicago World's Fair, where Ford's line went from ten thousand cars to two million, where Bell's telephone went from a curiosity in wealthy homes to a utility in every business, where IBM's PC went from a hobbyist's toy to a corporate necessity. **The moment where everything accelerates.**

And acceleration is the whole point, because the clock is the entire story. Watch the forty-year electricity cycle collapse: the telephone took roughly thirty years to become the default, the automobile about twenty, computing about fifteen, the internet about ten, mobile around five. Each shift compressed the timeline because each one inherited the infrastructure of all the shifts before it. The internet was faster than computing because computing already existed. Mobile was faster than the internet because the internet already existed. **AI is faster than all of them because it rides every layer ever laid: the data centers, the fiber, the cloud, the mobile devices, the global internet.** Every prior shift laid track for this one. The acceleration is not slowing. It is compounding.

AI is tracking at three to five years. The window opened November 30, 2022. As of mid-2026 we are roughly three and a half years in. Even the generous reading, that AI's window lasts half the internet's, closes it around 2028; the compressed reading closes it sooner. Both calculations converge on the same uncomfortable answer: **the decisive phase is the next twenty-four months.**

And understand exactly what that window measures. The telegraph operator who laughed at the telephone in 1876 had fifteen years before his industry died. The horse-and-buggy maker who dismissed Ford's line in 1908 had about a decade. The candlemaker who ignored electricity had maybe twenty years. **You might have three.** And those three years are not three years before AI starts touching your industry. **They are three years before the positions are locked**, before the finishers have captured the platforms, the ecosystems have hardened, the standards have congealed, the winners and losers have been decided. After that you are not positioning for the future. **You are scrambling to survive inside someone else's.**

The pattern is documented. The Sequence has never failed. Five stages. Inventors lose. Finishers win. Incumbents fall. Jobs compress. Value concentrates. And, the rule we bookmarked at the telephone, **even total dominance carries an expiration date.** AT&T ran American communications for more than a century and was broken up and effectively consumed by its own offspring; the brand survived, the original entity did not. No position is permanent, not even the biggest. The Sequence is impersonal. The only variable it cannot predict is you.

So here is the work, now, before the next page. Locate yourself. **Which stage is your industry in? Is AI still inventing in your field, or has the infrastructure already arrived?** Are you an inventor defending patents, pouring your best energy into protecting what you built yesterday? Or a finisher building the infrastructure that will make AI inevitable in your domain? Are you Ford in 1913, building the system that puts the product in everyone's hands, or the Wright Brothers in a courtroom while the world flies past? Are you IBM, certain the visible hardware is what matters, while someone else quietly takes the invisible platform underneath you? The pattern is not a theory; it is a historical fact, proven five times in this chapter alone. Five industries, five eras, five identical outcomes. The inventors who clung to their patents lost. The finishers who built infrastructure won. The incumbents who dismissed the shift disappeared. **Your only decision is which role you intend to play in the seventh turn, and you are deciding it right now, whether you admit it or not.**

In the next chapter we speed-run six more shifts (manufacturing, computing, internet, e-commerce, mobile, and cloud) to show you exactly how violently the timeline is compressing. Each one happened faster. Each one was bigger. Each one ran the identical Durable Sequence.

The acceleration is real. And it is about to hit you.

Key Takeaways

- The **Durable Sequence** is not a metaphor; it is a law with five stages in fixed order — *Invention → Infrastructure → Platformization → Proliferation → Consolidation* — proven across electricity, telephony, aviation, automobiles, and computing without a single exception.
- The **finisher** who builds the infrastructure and wins the standard captures the value; the inventor who proves the idea earns the footnote.

Edison, Vail, Boeing, Ford, and Gates all finished what someone else invented.

- Incumbents die by judging the new technology on the old technology's scoreboard, and the most rational analysis in the room is usually the fatal one. The telephone *was* worse, right up until it owned everything.
- The platform is almost always the **invisible layer**: Microsoft's software, not IBM's hardware. Find the layer everything in your industry is forced to run on; that is where the toll lives.
- Winning is not safety. Ford finished the automobile and was nearly finished by the **Success Trap**; AT&T ran communications for a century and was consumed by its own offspring. No position is permanent.
- The clock has compressed from forty years to **three-to-five**, opened November 30, 2022, roughly three and a half years in as of mid-2026, decisive phase through ~2028. A three-year roadmap is a surrender written in advance.

Finisher Move: Place your industry on the five stages this week: one coordinate, defended in writing. Then name the single piece of infrastructure or standard everyone in your field will be forced to route through, and decide whether you are building it or about to rent it.

Boardroom Question: *"What is the invisible platform layer in our industry, and who is quietly capturing it right now, while we are still admiring the visible one?"*

Sources & Currency Note

This book states facts as of mid-2026; specific figures move, the pattern does not. Stable historical anchors used here (Pearl Street Station, 1882; the Bell–Gray same-day patent filing, Feb 14, 1876; the AT&T divestiture, 1984, and SBC's 2005 acquisition; Kitty Hawk, Dec 17, 1903; Highland Park's moving line, 1913, and the $5 day announced Jan 5, 1914; IBM's MS-DOS licensing decision; ChatGPT's launch, Nov 30, 2022) are widely documented. Perishable figures are framed accordingly: NVIDIA as the first company to cross a $5 trillion valuation

(October 2025); 2026 hyperscaler AI capital expenditure on the order of ~$700 billion; the Microsoft–Constellation deal to restart Three Mile Island Unit 1 (announced Sept 2024, restart targeted for 2028, later accelerated toward 2027); ChatGPT weekly active users (more than 900M by May 2026 and still approaching ~1B, OpenAI); and the collapse in frontier-AI token pricing (a >90% fall over roughly three years — e.g., GPT-4-class blended pricing from the tens of dollars per million tokens at its 2023 debut to a few dollars by 2024–2026, with cheaper capable models far below that). Older industrial figures (electricity cost curves, Model T prices and market share, IBM's ~70% mainframe share and >$25B revenue, peak AT&T valuation) are presented as documented historical estimates and framed as ranges where period sources vary.

CHAPTER 2

SIX SHIFTS, ONE PATTERN

IN THIS CHAPTER the proof. The Durable Sequence didn't run once – it ran six times across two centuries, and each turn closed faster than the one before it. **Why it matters:** the same compression that took electricity forty years to become the default now gives AI three to five – the shortest window any generation has ever been handed. **What to do with it:** stop budgeting for AI as a multi-year initiative and start treating the next twenty-four months as the entire game.

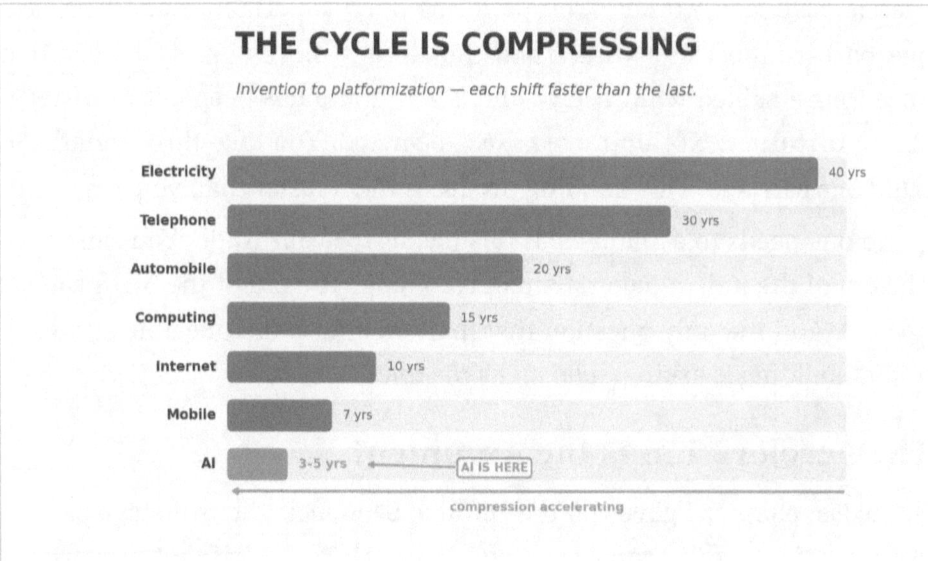

It is 1913, and you are a master carriage maker in Detroit.

You have spent your life building the finest horse-drawn carriages in Michigan. Your hands know the grain of every board. You can bend a steel frame by feel and shape a leather seat to any customer's spine. Each carriage takes you three months, and wealthy families pay handsomely for what only your hands can do.

A friend talks you into visiting Henry Ford's new Highland Park plant. You walk through the doors and the first thing that hits you is the smell. Machine oil and hot metal, sharp and acrid, nothing like the warm scent of wood shavings you have breathed your whole life. Then the sound: thousands of men working in unison, a rhythmic roar that comes up through the concrete and into your chest. You stand at the edge of the floor and watch something that makes your stomach drop. And you understand, with a clarity that tightens your throat, that you are not looking at a faster way to build carriages. You are looking at the end of everything you know.

Men stand shoulder to shoulder along a moving belt. The belt carries a chassis past each worker, who performs one task. Tighten one bolt, seat one part, connect one wire. Then the chassis moves on. The work that once took a skilled team twelve and a half hours has been cut to **ninety-three minutes.** Not improved. Recategorized. You take three months to build one carriage. This building produces them faster than you can count.

No one needs to explain what this means for your trade. You can see it. The era of the individual craftsman is closing. The era of the platform has opened. And the only question that matters now is the one almost no one in that building is asking: *who owns the line?*

The Acceleration Is the Argument

In the last chapter I gave you the Durable Sequence, the five-stage pattern every platform shift runs through: **Invention → Infrastructure → Platformization → Proliferation → Consolidation.** I proved it with deep cases. The inventor proves the idea and earns a footnote; the finisher builds the infrastructure everyone is forced to route through and owns the toll for a generation.

This chapter proves the second half of the thesis, the half that should change your calendar, not just your strategy. **The sequence never changes. The clock always shrinks.**

Electricity took roughly **forty years** to go from invention to default. The automobile, about **twenty.** Computing, **fifteen.** The internet, **ten.** Mobile and cloud, about **five each.**

AI gets **three to five.**

Read those numbers as a series, not a list. Each window is roughly half the one before it. Every generation gets less time to see the shift, adapt to it, and claim a position before the positions are assigned. The people who lived through the electrification of industry had four decades to find their footing. The people who lived through mobile had five years. You have been handed something shorter than that, and you have been handed it without the courtesy of an announcement.

The compression has a cause, and the cause is the reason it will not slow down: **every new platform inherits the infrastructure of all the platforms before it.** Electricity had to wire the streets. The internet had to lay the fiber. AI has to wire nothing. It propagates instantly through the billions of screens already in your customers' hands. There is no construction delay between capability and ubiquity anymore, which is exactly why the window is the narrowest in the history of capital.

Let me run six shifts past you in rapid succession. Watch for two things: the sequence is identical every time, and the clock halves every time. By the sixth, the argument should feel less like history and more like a countdown.

Shift One: Manufacturing — Ford Owns the Line

The automobile established the Durable Sequence in the last chapter. Here, take the timeline and the wreckage.

The car went from luxury to necessity in roughly twenty years, from fewer than one in a hundred American families to more than half. But the cars were never the point. The second-order effects dwarfed the industry that created them: suburbs, highways, consumer credit, the oil economy, fast food, shopping malls, an entire century reorganized around the assumption of personal mobility. **This is the signature of a platform shift. It does not add to the existing economy. It multiplies the**

whole economy by a factor no one forecasts, and it reroutes that multiplied value through whoever owns the new infrastructure.

The losers are not a metaphor. The **253 American automakers** operating in 1908 had collapsed to **44 by 1929**, with roughly four-fifths of all output running through just three names: Ford, General Motors, Chrysler. By the 1960s the survivors were those same three. A consolidation in the high nineties of a percent. The carriage maker standing at the edge of Ford's floor had perhaps twenty years before his trade was gone, and he was one of the lucky ones, because he could *see* the line. Most incumbents never get to watch their replacement run.

Ford himself would later become a casualty of his own success, nearly destroying his company by refusing to move past the Model T. That is the exact failure mode we name in the next chapter. The pattern does not spare the finisher who stops finishing.

Clock: invention to default — about twenty years.

Shift Two: Computing — The Standard Beats the Hardware

Before the personal computer, computing was a priesthood. A mainframe cost between one and five million dollars, lived in a refrigerated room, and was tended by specialists in white coats. You did not touch it. You handed your program to the priesthood on punched cards and waited hours or days. In 1970 there were on the order of fifty thousand computers on Earth, and the idea that an ordinary person might own one was as absurd as owning a jetliner.

The invention phase detonated in 1975, when the Altair 8800 hit the cover of *Popular Electronics*: a $439 kit with no keyboard, no screen, and no software, programmed by flipping switches and reading blinking lights. Barely usable. But it proved the thing was possible, and within five years more than a hundred incompatible makers were selling personal computers. The Apple II, the Commodore PET, the TRS-80, each with its own operating system, its own software, its own island of loyalists. Pure proliferation. A hundred visions, no standard, nothing compatible with anything.

Then IBM entered in 1981 and made two decisions that looked reasonable and proved fatal.

First: open architecture. IBM published the full technical spec, assuming it would seed a peripheral ecosystem. What it seeded was clones. Compaq, Dell, and others building cheaper machines that ran identically. IBM gave away the hardware moat in a single document.

Second, and this is the one to carve into the wall: IBM outsourced the operating system to a small Seattle company and let it keep the rights. IBM wanted to buy MS-DOS outright. Bill Gates refused, insisting on a license that let Microsoft sell the same OS to anyone. IBM agreed, certain no other PC maker would ever matter. They were the most powerful technology company on Earth. Who could possibly challenge them?

Every clone maker needed an operating system, and Microsoft sold it to all of them. In one stroke, the standard detached from the hardware, and Microsoft owned the standard. Every manufacturer needed Windows. Every developer wrote for Windows. Every user learned Windows. The flywheel of network effects spun until, by the early 1990s, competing with Windows was like launching a rival telephone network after everyone already had a phone. Microsoft's revenue climbed from roughly $140 million in 1985 to more than $20 billion by 2000, and its market value from a few hundred million to the high hundreds of billions. Apple, meanwhile, fell to single digits of market share and came within **less than ninety days of running out of cash in 1997**, saved, with exquisite irony, by a $150 million lifeline from Microsoft. And IBM, the company that *created* the PC standard, eventually walked out of the PC business entirely and sold the division to Lenovo.

The computing shift hands AI two laws. The obvious one: **whoever owns the platform standard captures the value.** IBM invented the standard and lost it; Microsoft licensed the OS and took the industry. The subtler, more dangerous one: **the platform winner does not have to own the best technology.** MS-DOS was widely considered inferior to the Macintosh's operating system. It did not matter. Microsoft won on distribution and network effects, not elegance. The best product and the winning platform are rarely the same thing, a sentence worth taping to the wall of every AI lab racing for benchmark supremacy right now.

Today the same contest is live. As of mid-2026, OpenAI, Anthropic, Google, and Meta are all maneuvering to set the standard for how AI is accessed, deployed, and embedded. The company that becomes the *Microsoft of AI*, the default everyone else is forced to build on, will capture a disproportionate share of the value regardless of who ships the most impressive model. If computing is any guide, the winner will not necessarily own the smartest system. It will own the distribution: the one that makes itself indispensable to the most developers and enterprises, and locks in switching costs so high that leaving becomes unthinkable.

Clock: invention to default — about fifteen years.

Shift Three: The Internet — When Even the Reigning Platform Misreads the Next One

The internet's invention phase runs back to ARPANET in 1969, a handful of university machines wired together. For two decades it stayed an academic curiosity. By the late 1980s perhaps a hundred thousand computers were connected, almost all in research labs. Ordinary people had never heard of it.

The infrastructure phase accelerated in the early 1990s: Tim Berners-Lee's World Wide Web in 1991 made the network navigable; the Mosaic browser made it usable in 1993; Cisco built the routers and switches that made it fast; providers like AOL made connecting trivial.

And here is the detail most strategists skip, because it is the most threatening one: the reigning platform owner missed it.

In 1991, Bill Gates's internal "nightmare" memo was not about the internet at all. It was about IBM and Novell, the last war. The web exploded anyway. By early 1994 Microsoft was still circulating internal "internet primers" explaining what this network was becoming, and Gates told a senior offsite that the internet "mania" had taken him by surprise. Millions of users were standardizing on software Microsoft had neither designed nor controlled. **You can be the dominant platform of one era and still be blindsided by the next, precisely because you are optimized for the last one.** Hold that thought against every incumbent reassuring its board that AI is "an extension of our existing stack."

The platformization phase hit in the mid-to-late 1990s. Access costs collapsed from $25–50 a month for dial-up to a flat line item on bills people already paid once broadband rode in. Users went from a few million in 1993 to over a billion by 2005. And the ecosystem erupted: search, email, news, forums, commerce, social. The value pooled, as always, in the infrastructure and the platforms. **Cisco, which made the routers, became the most valuable company in the world in March 2000**, worth more than half a trillion dollars at its peak. Google organized the web until its name became a verb. Amazon built the cloud half the internet would later run on.

The losers were every incumbent who mistook a platform shift for a passing fad, and the destruction was not gradual erosion. It was structural annihilation of business models that had felt permanent.

Start with **newspapers.** For a century they ran one of the most profitable models in capitalism: control the local information (council meetings, ball scores, obituaries, classifieds) and you control local advertising, at thirty to forty percent margins, decade after decade. The model felt as fixed as gravity. Then Craigslist, started by one man and still run by a few dozen people, vaporized classified advertising, billions in annual revenue, by making listings *free*. Google and Facebook took the display advertising that remained by offering targeting and measurement newspapers could never match. Hundreds of papers closed inside a decade. Not because people stopped wanting news, but because the platform owned the distribution and no longer needed the newspaper to reach the reader.

The **music industry** suffered the parallel catastrophe. US recorded-music revenue fell from a peak of roughly $14.6 billion in 1999 to under $7 billion by 2014, more than half the industry, gone. Napster, written by an eighteen-year-old in a dorm room, did more damage to recorded music than any competitor in its history. One teenager with an idea and a computer dismantled a global distribution empire built over a century. The industry's response was to spend years suing its own customers and lobbying Congress instead of adapting, the identical error the telegraph companies made when the telephone arrived. Right about the principle. Catastrophically wrong about the trajectory. **Being right about the principle is no defense when the trajectory is against you.**

And the brick-and-mortar incumbents who dismissed e-commerce simply disappeared into it. Borders filed for bankruptcy. Circuit City vanished. Sears, the company that built mail-order retail into an empire, that pioneered remote commerce a century earlier, could not adapt to its own digital successor and collapsed.

Clock: invention to default — about ten years (web in 1991 to mass default by 2001).

Shift Four: E-Commerce — The Difference Between a Website and a System

In the mid-1990s, hundreds of entrepreneurs proved e-commerce was possible. Pets.com would ship pet food. Webvan, groceries. eToys, toys. Kozmo, anything in an hour. They raised billions, ran Super Bowl ads, and described revolutions. Almost all of them were bankrupt by the dot-com crash of 2000–2001.

They made the **inventor's mistake**: they proved the concept and never built the infrastructure that would make it inevitable. They had websites and no logistics. Customers and no fulfillment. Vision and no platform.

Amazon survived because Jeff Bezos refused to confuse the two. E-commerce, he understood, was never about putting a catalog online. It was about building the logistics spine that would make online commerce unavoidable.

This deserves a hard pause, because the Wall Street reaction to Amazon's strategy is almost word-for-word the reaction to AI infrastructure spending today. Every time Microsoft or Google announces tens of billions in data-center capex, analysts ask whether the spend can possibly be justified. They are reading from the script written about Amazon in 1999.

Amazon lost money for years, by design. Bezos poured every dollar back into warehouses, delivery, and systems while competitors bragged about margins. The street was merciless. They called it *Amazon.bomb*. In June 2000 a prominent Lehman Brothers analyst published a report arguing the company would run out of cash within four quarters. The stock fell from over $100 to under $10, roughly ninety percent of its value, erased, because Wall Street, for all its sophistication, could not tell the difference between *losing money* and *investing in infrastructure*. They saw red ink and read failure. Bezos saw red ink and read a moat his competitors could not cross in time.

Pets.com built a website. Amazon built a system. There is an ocean between those sentences. Amazon built warehouses within two days' shipping of most of the US population. Software that predicted purchases before customers knew they wanted them. A reviews layer that manufactured trust at scale, solving the founding problem of buying from strangers online. And then the move that changed the company's nature: Fulfillment by Amazon, which let any seller on Earth rent Amazon's logistics. Amazon stopped competing with other retailers and started competing with the very idea of building your own infrastructure. If you sell handmade candles, why would you build a warehouse, a shipping network, payment processing, a trust system, and a discovery engine when Amazon rents you all of it for a percentage? Millions of sellers reached that rational answer at once. By the mid-2020s, third-party sellers accounted for well over half of everything sold on Amazon. The retailer had become the operating system for commerce.

Then came the move that detached Amazon from retail entirely. The cloud infrastructure it had built to run *itself* (the servers, storage, databases, networking) turned out to be valuable to every other company alive. Amazon essentially said: *we built this for us; want to rent it?* Netflix, Airbnb, Slack, NASA, and eventually half the internet said yes. AWS became the invisible floor under the digital economy. When you stream a film or hail a ride, you are very likely running on Amazon's machines without knowing it.

The analysts who mocked the spending in 2000 were measuring the wrong thing. They counted quarterly profit while Bezos accumulated infrastructure. He was playing a different game than the one they were scoring. By the time the street understood what had been built, the company was worth trillions and the early critics had become case studies in the cost of short-term thinking during a platform shift. **Nobody remembers the sock-puppet mascot except as a punchline. Everyone routes through Amazon.** Read that as the prophecy it is.

Clock: invention to default — about seven years (early e-commerce in 1995 to Amazon's dominance by 2002–2003).

Shift Five: Mobile — A New Platform Hiding Inside a Worse Phone

Mobile computing existed for years before the iPhone. BlackBerry owned business email. Nokia sold roughly **forty percent** of all the phones on Earth. Palm had its faithful. Windows Mobile ran on dozens of handsets. The industry was established, profitable, and apparently stable.

Then on January 9, 2007, Steve Jobs walked on stage and said Apple was going to reinvent the phone.

The incumbents laughed. Microsoft's CEO mocked the price and the missing keyboard and declared it would never sell to business. Nokia's executives filed it under "toy for Apple fans." BlackBerry's leadership noted, correctly, that it could not even do proper email.

They were all right about the facts. The iPhone was expensive, keyboard-less, and worse at email than the devices it challenged. It lost on every metric the existing industry knew how to measure, which is exactly why the incumbents could not see it. The iPhone was not competing on those metrics. It was a new platform wearing the costume of a worse phone. The App Store, launched in 2008, let any developer on Earth build and sell software to every iPhone owner, democratizing mobile software the way the web had democratized publishing. A student in Bangalore could ship an app to a banker in Manhattan inside a week. Within two years there were hundreds of thousands of apps; within five, more than a million. Entire companies (Instagram, WhatsApp, Snapchat) existed only because the platform did. The iPhone was never a better phone. It was a pocket computer that happened to make calls, and that distinction decided who lived.

The collapse was the fastest the technology world had ever witnessed. Nokia fell from forty percent of the global market to near-zero in roughly five years. BlackBerry went from owning enterprise mobile to irrelevance in four. Microsoft spent **more than eight billion dollars** trying to force Windows Phone into existence and failed completely. We dissect *why* these giants fell, the specific mechanisms of the **Success Trap**, in the next chapter. For now, take only the verdict: **platform shifts do not respect incumbency.** They do not care how much money you hold or how thoroughly you won the last era. If that was true for Nokia, BlackBerry, and Microsoft, sit with what it means for today's incumbents standing in front of AI.

Consolidation came on schedule. Apple captured the platform through a closed ecosystem; Google captured it through Android's open-source giveaway, the same play Microsoft ran with MS-DOS a generation earlier. Together they now hold **on the order of ninety-nine percent** of the mobile operating-system market. Two companies. And the platform manufactured entire industries that could not have existed before it: ridesharing, food delivery, mobile-first social media, the gig economy, mobile banking and payments, every one of them built on the single assumption that everyone now carries a computer in their pocket.

Clock: invention to default – about five years (iPhone in 2007 to platform dominance by 2012).

Shift Six: Cloud – The Most Important Shift Almost No One Watched

Cloud computing may be the most consequential platform shift that happened in plain sight while the public looked elsewhere.

Before the cloud, every company ran its own servers, and the reality was punishing. To launch anything you bought physical machines costing thousands each, racked them in a data center that billed you for space, power, and cooling, hired administrators to nurse them around the clock, and (the expensive part) *guessed* your capacity months ahead. Guess low and a traffic spike crashed you; so everyone guessed high and massively over-provisioned. A startup could burn half its venture capital on infrastructure before writing a line of product.

Amazon launched AWS in 2006 with two modest-looking products: storage and computing power you rented by the hour instead of buying. Most established technology companies ignored it or dismissed it as a retailer wandering out of its lane. *Who would trust their business-critical systems to Amazon's servers?*

Everyone. Eventually, everyone. AWS deleted the infrastructure burden outright. Instead of months and millions, you launched in hours with no upfront cost and paid only for what you used; spikes scaled up automatically and troughs scaled down. The platformization was quiet and total: Netflix, Airbnb, Slack, thousands of companies, all built on Amazon's cloud. As of mid-2026 the three largest providers control roughly **two-thirds of the global market**: AWS around thirty percent, Microsoft Azure roughly a quarter, Google Cloud in the low teens. Most of the internet runs on infrastructure owned by three companies, and the margins are extraordinary: AWS alone throws off more operating profit than all of Amazon's retail business combined.

The cloud teaches the lesson most directly aimed at this moment: **the decisive infrastructure battles happen in the background, invisible to the people watching the foreground.** While the press and the public fixate on consumer-facing AI (the chatbots, the image generators, the viral demos) the real contest is in the layer nobody photographs: NVIDIA designing the chips, the hyperscalers providing the compute, the holders of the training data and the model-serving pipes. Those are the companies quietly deciding who owns the AI platform for the next generation, exactly as AWS became the backbone of the internet while everyone was busy watching the apps.

Clock: invention to default — about five years (AWS in 2006 to dominant cloud platform by 2011).

The Pattern Underneath the Pattern

Put the six on a single line and the compression stops being a claim and becomes a measurement. Electricity, forty years. Manufacturing, twenty. Computing, fifteen. The internet, ten. Mobile and cloud, five each. Every window roughly half the last. **AI: three to five**, and the invention phase is already finished, the infrastructure is being poured in public right now, and the platformization has begun.

Every one of the six ran the Durable Sequence in order. Every one saw the finisher capture the value while the inventor became a footnote. Every one destroyed incumbents who held every conceivable advantage. Every one generated second-order effects worth more than the original technology. And every one closed faster than the one before.

But there is a second pattern riding underneath the timeline, and it is even more consistent: **the consolidation.** It is so reliable across all six shifts it should be treated as a law of the physics of platforms.

In manufacturing, 253 automakers became 44 and then three. In computing, dozens of operating systems and hundreds of hardware makers shook out to two operating systems that mattered, one of which, Windows, held over ninety percent of the market for two decades. On the internet, thousands of search engines (AltaVista, Lycos, Excite, Yahoo) collapsed to one that held the overwhelming majority of search. In mobile, dozens of operating systems narrowed to two holding nearly the entire market. In cloud, hundreds of hosting providers reduced to three controlling two-thirds of the world's infrastructure.

The shape is always identical: **explosive proliferation, then brutal consolidation.** Early on, barriers are low and excitement is high and hundreds of players flood in; investors fund anything that moves; the press calls it democratic, open, infinite. Then network effects ignite. The platforms that reach critical mass become self-reinforcing (more users pull more developers pull more users) and the platforms that don't discover the cruelest rule in the system: **in a platform market, second place is the same as last.** Within a few years the hundreds are a handful, and the rest become customers of the survivors or footnotes in a case study. The consolidation phase is where the real fortunes are made, and where most participants lose everything they put in.

AI will run this exact arc. As of mid-2026 we are deep in proliferation: hundreds of startups, a dozen serious foundation models from OpenAI and Anthropic and Google and Meta and others, thousands of applications, money pouring in from every direction. It *feels* wide open. It feels like anyone can win. That is precisely how computing felt in 1980, how the internet felt in 1996, how mobile felt in 2008, months before the door closed. Within three to five years the consolidation will begin in earnest: a small number of companies will own the platform infrastructure everyone else depends on (the foundation models, the compute, the developer ecosystems, the distribution). The rest will rent.

There is no middle ground in platform economics. You own the platform or you pay to use someone else's.

> SIGNAL TO WATCH The consolidation never fails. 253 automakers became 3. Dozens of operating systems became 2. Thousands of search engines became 1. If you are not building toward platform scale, owning a piece of infrastructure others must route through, you are, by default, building toward becoming someone's customer.

If six shifts in a hundred years still reads like history rather than prophecy, let me show you where the curve actually arrives.

2:00 A.M. in Oshino

LIGHTS-OUT PRODUCTION

A fully automated hall — no human shift, lit only by indicator lights.

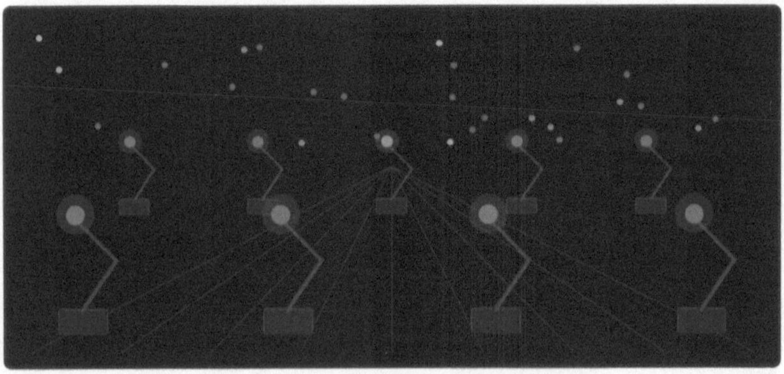

The factory runs in the dark. The only thing it doesn't need is us.

It is two in the morning at a FANUC factory in Oshino, Japan, and the lights are off.

They have been off for years. There is no reason to turn them on. There is no one inside to see. In the dark, six-axis robotic arms swing through precise arcs, grasping components, welding joints, assembling boards to tolerances measured in microns. Conveyors hum at frequencies tuned for efficiency, not for human comfort. Pneumatic pistons fire in rhythms no foreman ever set. Inspection cameras flash in infrared no human eye could register.

Robots are building other robots. They have been doing it here since 2001, around fifty per shift, and the line can run unmanned for weeks at a stretch. No shift changes. No breaks. No sick days, no vacation requests, no grievances. The defect rate is essentially zero, better than any human line has ever achieved, anywhere, in the entire history of manufacturing.

One security guard walks the perimeter. He is the only human in the building. His footsteps echo off concrete poured for hundreds of workers who will never arrive. He checks his phone, scrolls a message from his wife, and glances up at the arms swinging through their arcs in the dark.

The factory does not notice him. It does not notice anything. It does not need to. It simply produces, perfectly, endlessly, in the dark.

This is what the far end of the acceleration curve looks like. Not a speculation, not a demo. A factory that has run lights-out for years, building the machines that build the next generation of machines, in a loop of automated reproduction that no longer requires human hands, eyes, or presence.

The carriage maker in 1913 watched men build cars faster than he could build a carriage, and he had about twenty years to find his next move. The guard in Oshino watches machines build machines in the dark. The distance between those two scenes is the distance the curve has already traveled, and it is still accelerating.

The carriage maker had twenty years. **You have three.** Spend them building the line, or spend them standing at the edge of someone else's, watching.

Key Takeaways

- The Durable Sequence didn't run once — it ran six times (manufacturing, computing, internet, e-commerce, mobile, cloud), and **every turn closed faster than the last:** ~40 years for electricity down to ~5 each for mobile and cloud. **AI gets three to five** because it inherits all prior infrastructure and propagates instantly through screens already in hand.
- In every shift the **finisher who owned the platform** captured the value, and **the best technology rarely won** — MS-DOS beat the Macintosh OS on distribution, not quality. Owning the standard beats owning the smartest product.
- The **consolidation is a law, not a tendency:** 253 automakers became 3; dozens of operating systems became 2; thousands of search engines became 1. Second place in a platform market is the same as last.
- The most decisive battles happen **in the invisible infrastructure layer** — chips, compute, data, distribution — not in the consumer demos everyone is watching.
- A website is not a system. **Pets.com built a website; Amazon built infrastructure** — and the market that mocked the spending became its case study. Wall Street's reaction to AI capex today is the 1999 script, reprinted.

Finisher Move: Find your industry's last true platform disruption, measure how many years it took to play out, then divide by four or five. That number, not your inherited planning horizon, is how long you have to claim a position before the consolidation assigns one to you.

Boardroom Question: *"If our competitive window is three years instead of fifteen, what do we stop doing tomorrow — and what piece of infrastructure do we start owning?"*

Sources & Currency Note

This book states facts as of mid-2026; specific figures move, the pattern does not. Key data points in this chapter: Ford's Highland Park moving line cutting chassis assembly from ~12.5 hours to ~93 minutes (1913); the *consolidation of US automakers from 253 (1908) to 44 (1929), ~80% of output through Ford/GM/Chrysler; Apple within "less than 90 days of cash" in 1997 and the $150M Microsoft investment; Amazon's stock falling from over $100 (1999) to under $10 (2001) — roughly 90% — and the prominent Lehman Brothers analyst's "four quarters" report (June 2000); Cisco as the world's most valuable company in March 2000 (>$500B peak); US recorded-music revenue down from ~$14.6B (1999) to <$7B (2014); the Apple/Google mobile-OS duopoly (~99% combined); the cloud "big three" (~two-thirds of market — AWS ~30%, Azure ~25%, Google Cloud low-teens, mid-2026); and FANUC's Oshino lights-out plant (robots building robots since 2001, ~50/shift, unmanned for weeks). Compression curve and consolidation framing are the book's own analysis of public history. Where a figure could age, it is dated or given as a range.*

CHAPTER 3

THE SUCCESS TRAP

> **IN THIS CHAPTER** the five-mechanism machine that converts a market leader's every strength into the cause of its death during a platform shift — and the brutal 2026 receipts proving the machine is running inside companies right now, including the ones selling AI. **Why it matters:** if you are winning today, the trap is not a risk you might encounter; it is a system already operating inside your walls, and it is invisible precisely because it looks like good judgment. **What to do with it:** run the five-mechanism diagnostic on your own organization this quarter — before a competitor, or a journalist, runs it on your obituary.

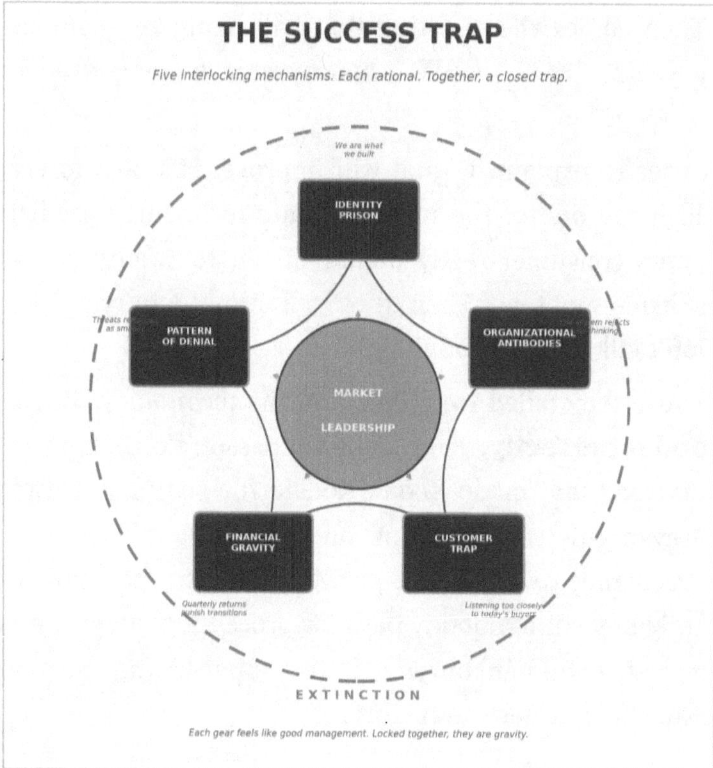

It is early 1976, and Steven Sasson is nervous.

The twenty-five-year-old engineer walks a long corridor inside Kodak's Rochester headquarters carrying a device that looks like a toaster bolted to a circuit board. It weighs eight pounds. It captures images at 0.01 megapixels: ten thousand pixels, a grainy hundred-by-hundred black-and-white square. It takes roughly twenty-three seconds to write one photograph onto a cassette tape, and you need a separate playback rig wired to a television to see the result. He is about to set this thing on a conference table in front of the executives who run the most dominant imaging company on Earth.

"This takes photos without film," he says.

They pass it around. They ask sharp technical questions, and he answers them with the precision of an engineer who built the thing from nothing. Then comes the question that will haunt the company for the next forty years: *"Why would anyone want to look at photos on a television screen?"*

Sasson tries to explain. Digital will improve. His own forecast, which history will judge one of the most accurate technology predictions ever recorded, puts consumer-ready digital fifteen to twenty years out. The executives listen politely. Then they tell him to bury it. Don't show anyone. Don't talk about it. Don't publish.

Not because they failed to understand the technology. **Because they understood it perfectly.** They looked at Sasson's prototype and saw the end of everything that made Kodak Kodak: the sixty-percent margins on film, the seventy-percent margins on processing, the razor-and-blade engine of recurring revenue that printed billions. They saw a future in which nobody buys film, nobody pays for processing, nobody needs them. So they buried it, and in burying it they opened the most instructive corporate suicide in American history.

The part most readers miss is the only part that matters. Sasson did not lose to people who couldn't see the future. He lost to people who saw it more clearly than he did and chose the present anyway. He went back to his lab and kept working at Kodak for decades, watching from the inside as his forecast came true almost exactly on schedule, watching competitors build what Kodak already owned the patents to, watching a hundred and forty-five thousand jobs evaporate because the men in that room decided comfort now was worth more than survival later. That is the Success Trap at its most human. Not stupidity. The deliberate, rational, well-defended choice to protect what is working until what is working is gone.

I am going to name the machine that made that choice for them: five mechanisms that operate together with the reliability of gravity. Name the machine and you stop being its casualty. Miss it, and you become its next case study.

The Most Dangerous Position in Business

Start with the paradox that should cost you sleep: **the most dangerous position in any platform shift is the lead.**

Leaders hold everything. Resources that dwarf the upstarts. Talent the upstarts can't afford. Customers who trust them on reputation alone. Brands worth billions, distribution built over decades. By every visible measure they should own the next era as completely as they own this one.

They almost always lose. Not sometimes. *Almost always.* And this is not a bug in the system; it is the system's defining feature. Success does not prepare an organization for transformation. Success prevents it. The exact capabilities that won the last era harden into the chains that drown you in the next.

Run the roll call and watch the law execute. Western Union dominated global communication and had first claim on Bell's telephone patents, then declined them as an inferior toy. The gas companies lit the industrial world on monopolistic franchises, and electricity erased them inside a generation. Nokia sold forty percent of the phones on the planet, and the iPhone reduced it to a footnote in roughly five years. Yahoo was the front door to the internet. It passed on buying Google twice, then turned down Microsoft's roughly forty-four-billion-dollar offer, before selling its core to

Verizon years later for around a tenth of that. Blockbuster had the chance to buy Netflix for about fifty million dollars and laughed the offer out of the room.

Every single time, the same five moves in the same order: the dominant company sees the threat, dismisses it as inferior, retreats to protect the existing business, deploys a half-measure to look responsive, and dies. The question is not *whether* leaders fail. It is *why* the smartest, richest, most capable organizations on Earth fail with such grim consistency.

Before I answer, I have to close an escape hatch your mind is already reaching for. Kodak, Blockbuster, Nokia. These feel like history. Sepia. Lessons safely learned. You are reading them thinking: *that was them; we're sharper; we'd see it coming*.

You wouldn't. And I can prove it, because the same machine is turning right now, in a corner office you could walk into this afternoon.

The Partner Who Put Down the Phone

It is a Monday in Midtown Manhattan, forty-third floor of a BigLaw firm. The managing partner, call him David, is reading a weekend briefing on JPMorgan Chase's contract-intelligence system, **COIN**. As reported in 2017, COIN reviewed roughly **12,000 commercial credit agreements**, work that had consumed about **360,000 hours of lawyer and loan-officer time every year**, and did it in seconds. Three hundred sixty thousand hours. Call it a hundred and eighty full-time people, year-round, reading contracts, checking clauses, flagging anomalies. The exact work his associates do in the bullpen outside his door.

He looks up. Through the glass he can see forty-seven associates at their desks. They bill, on average, around five hundred dollars an hour, roughly eighteen hundred hours a year. He does the arithmetic he does not want to do: that is on the order of forty million dollars in annual revenue, from precisely the kind of work a bank just automated to near-zero.

He picks up the phone to call the managing committee. Then he puts it down.

Because what is the sentence? *We need to fire half the bullpen because a bank built a robot that reads contracts faster than our associates can?* The firm's entire economic model rests on a single assumption: recruit the best graduates, train them three years, bill them at five hundred while paying them a fraction of it, all on the premise that this work requires expensive human intelligence and always will. COIN just falsified that assumption in production, at one of the largest financial institutions on Earth. So David slides the memo into a drawer. He goes to the partners' lunch. He doesn't mention it that day, or the next, or the one after.

That is the Success Trap migrating from a 1976 conference room to a 2026 corner office without changing a single gear. The partner sees the future. He runs the math correctly. And he chooses not to act, not from stupidity, but because acting means dismantling the very machine that made him rich. The trap was never something that happened to other people in another decade. It is happening, right now, to the most capable person in the building.

Why? Five mechanisms, working together like a machine engineered to destroy market leaders. Learn them and you will see them turning in your own company, and your own career, before you finish the chapter.

Mechanism One — The Identity Prison

The first law: an organization defends what it *is* harder than what it earns.

Kodak was not a company that happened to sell film. It was *a film company*. That identity ran through every hire, every budget line, every hallway conversation in Rochester, where employees spoke of having "yellow blood." Asking Kodak to become a digital company was not a strategy question. It was an identity transplant, asking an institution to become someone else.

Blockbuster was not a company that rented videos. It was *a store company*. The nine thousand storefronts were not distribution points; they were the embodiment of the thing itself. Strip away the stores and the question that froze the boardroom was unanswerable: *then what is Blockbuster?*

Nokia was *a phone company,* and the best one on Earth. Its engineers were the finest phone engineers alive. But the iPhone was not a better phone; it was a pocket computer that happened to make calls, and a phone company cannot make the leap to platform company without ceasing to be itself. Nokia kept building superb phones while Apple and Google built the ground those phones would have to stand on. The market stopped needing phone engineers. It needed platform engineers. Nokia had a surplus of exactly the wrong genius.

Identity prevents evolution. When the answer to *"who are we?"* is fused to *"what we have always made,"* you will choose to die as yourself before you will agree to live as something else. So here is the diagnostic, and you have three seconds to answer it: **if the technology that defines your industry vanished tomorrow, would your company still know what it was? Would *you* still know who you are professionally?** If the answer takes longer than three seconds, you are already inside an identity prison. You simply haven't found the walls yet.

Mechanism Two — Organizational Antibodies

The second law: a successful organization grows an immune system whose sole purpose is to kill anything that threatens the core, and it cannot tell a cure from a disease.

Watch the antibodies fire. It happens the same way every time, and every voice in the room is *correct.*

Sales comes back from the field: "I talked to our biggest accounts. They don't want this. They're happy. This would upset them." True. Current customers genuinely do not want disruption.

Finance runs the model: "Our core does sixty-percent gross margins. This new line does fifteen at best. It would crater profitability and the stock." Accurate. The math is airtight.

Operations weighs in: "Our entire infrastructure is built for the current model. Different supply chains, different capabilities, different everything." Honest.

Marketing objects: "This muddies a brand we spent decades building." They have a point.

Legal raises a hand: "The regulatory and liability exposure is unquantified. The risk is enormous." Doing their job.

HR delivers the finishing blow: "We don't have the people for this. Our whole workforce is optimized for what we do today."

Each department, reasoning impeccably from its own vantage, contributes one clean cut. The innovation dies of a thousand rational incisions, murdered by the very capabilities that made the company great. No villain stood up and said *let's destroy our future.* Everyone did their job *well,* and in doing it well they killed the company. This is why the Success Trap is not a leadership failure you can fire your way out of. It is structural. Swap every executive and the outcome holds, because the antibodies live in the organization's DNA, not in any one person's decision.

> SUCCESS TRAP WARNING Every department killed the innovation rationally. Sales said the customers don't want it. Finance said the margins are terrible. Operations said the infrastructure doesn't fit. Nobody was wrong, and that is exactly why the company died. Rational resistance is the deadliest form of denial, because you cannot argue a person out of a position they reached by being right.

Mechanism Three — The Customer Trap

The third law: during a platform shift, your best customers become your worst strategic advisors, and following them is rational, responsible, and fatal.

Your strongest accounts do not want the new thing. They are satisfied with the current solution; they have invested in it, trained their people, integrated their systems, built their workflows around it. For them, change is pure cost and the future looks strictly worse than the present. This sets a perfect trap: listen to them and you miss the shift; ignore them and you lose this year's revenue. Most leaders choose to listen, because listening feels like discipline. It leads straight off the cliff.

Kodak's professional photographers did not want digital in the 1990s; film genuinely was superior for their work. Kodak listened, protected film, and missed the consumer revolution entirely. By the time the pros switched, consumers had already left. Blockbuster's loyal members liked the stores, the browsing, the Friday-night ritual; Blockbuster listened and defended the experience while Netflix built the replacement. IBM's enterprise customers did not want personal computers, since the PC was underpowered for serious work, so IBM listened and handed the entire platform to Microsoft and Intel.

In every case the company did the textbook-correct thing. It was responsive. It was customer-focused. It delivered exactly what its customers asked for. And it died for it. The lesson is not that customers are foolish. It is that **existing customers can only describe what**

they want inside the current paradigm. They cannot tell you what they'll want in the next one, because in the next one, they will not be your customers at all.

> **FINISHER INSIGHT** Your best customers want the status quo *improved,* never *replaced.* Kodak's photographers begged for richer film. Netflix's DVD subscribers never once asked for streaming. The signal you need does not come from the people paying you today. It comes from the people not yet talking to you, who are, at this moment, talking to whoever is going to take your market.

Mechanism Four — The Financial Gravity Well

The fourth law: the numbers always argue against transformation, and the tragedy is that the numbers are right.

Your current business throws off high returns on predictable cash flows. You can model it to the decimal. The new technology offers lower returns wrapped in massive uncertainty. Any honest net-present-value calculation says stay. The CFO runs it, presents it, and the board blocks the transformation. The spreadsheet has spoken, and the spreadsheet is not lying.

The margin trap is crueler still. Your core may run forty to seventy percent gross margins, the compounded payoff of years of scale and optimization. The new model offers five to fifteen at best. Moving means voluntarily destroying your own profitability; Wall Street punishes the stock, the board replaces the CEO, and the company retreats to its core and marches toward bankruptcy with *improving quarterly margins all the way down.* And over all of it hangs the quarterly-earnings prison: a public company must hit its number every ninety days, which means it structurally cannot fund a transformation that pays off in years, cannot absorb the losses reinvention demands, cannot stand in front of analysts and explain why it is dismantling a profitable business to chase an uncertain one.

Kodak was generating billions in film revenue at sixty and seventy points. Digital offered a fraction of that with no recurring tail. Every analysis said protect film. Every quarterly report demanded film profits continue. And every quarter of film profits walked the company one step closer to the grave. The numbers didn't lie. **They simply could not see past the edge of the model that produced them.** A number that cannot see past its own assumptions is not analysis. It is gravity.

Mechanism Five — The Pattern of Denial

The fifth law: when the threat finally becomes undeniable, leaders run a three-act script (dismiss, half-measure, panic) with the predictability of a stage play.

Act one: dismissal. "It's a toy." "The quality isn't there." "Our customers don't want it." "We have time." Blockbuster's CEO called Netflix "a niche player." Nokia's leadership judged the iPhone "not a threat." The Microsoft CEO you met in Chapter 2 mocked the iPhone's price and missing keyboard on camera. They all saw it. They were all intelligent and experienced. They dismissed it anyway, because dismissal is psychologically cheaper than transformation.

Act two: the half-measure. When dismissal fails, the company does *something,* and something is worse than nothing. Kodak built digital cameras and deliberately crippled them to protect film: overpriced, underpowered, half-marketed. Blockbuster launched an online service and starved it to protect the stores. Nokia bolted touchscreens onto devices that still defaulted to physical keyboards. Every half-measure failed for the same reason: **platform shifts are binary.** You are on the new platform or you are not; there is no partial migration, no comfortable middle. But half-measures *feel* like progress while quietly avoiding the only choice that matters. They are corporate procrastination wearing the costume of strategy.

Act three: the panic pivot, too late. When customers are openly fleeing, the desperate lunge arrives after the window has shut. Kodak declared bankruptcy in 2012 and pivoted to printing, just as the world stopped printing. Blockbuster filed in 2010 and reached for streaming after Netflix had already crossed twenty million subscribers and was accelerating. Nokia sold itself for scrap. BlackBerry stopped making phones at all. By the time leadership finally moves, the insurgents own the platform, the standards have set, and the company is no longer a competitor. It is a slide in a business-school deck.

The Dabbler's Receipt

Most books about disruption stop here, with dead companies safely in the past tense, and a quiet implication that the danger now is *missing* AI. That framing is a trap of its own, and 2026 has already filed the receipts to prove it.

Recall the three types from the opening of this book: roughly ninety percent **Dismissers** who wait for proof until the proof is an obituary, nine percent **Dabblers** who run pilots and appoint a Chief AI Officer but never let AI touch a profitable workflow, and one percent **Finishers** who rebuild around the new platform and capture nearly all the value. The Dabbler is the most dangerous of the three, because dabbling *feels* like motion. And the Pattern of Denial has a Dabbler-specific failure mode the older case studies don't fully show: the half-measure that doesn't just stall. It costs *more* than doing nothing at all.

Read the 2026 ledger.

The arms dealers admit the ammunition costs more than the soldiers. NVIDIA's own VP of applied deep learning, Bryan Catanzaro, reportedly said it plainly: *"the cost of compute is far beyond the costs of the employees."* AI, for his team, is more expensive than the humans.

Uber reportedly burned through its *entire full-year 2026 AI budget by April,* with roughly 84 to 95 percent of its engineers using AI tools. And the single most telling data point of the year: **Microsoft,** among the most committed AI bulls on the planet, reportedly moved in 2026 to cut thousands of its own engineers' **Claude Code** licenses (cutoff June 30), shifting them to its in-house tooling, after per-engineer costs ran **$500 to $2,000 a month.** Note carefully: that is not a company that decided AI doesn't work. It is a company that found AI bolted onto its existing workflow cost more than the work it was meant to accelerate. (For the record: Claude is Anthropic's; Microsoft's large stake is in OpenAI, a separate firm, which only sharpens the irony. The most AI-invested company alive pulled an AI tool on cost.)

It scales to the absurd. As reported in 2026, by one account, an unnamed enterprise **accidentally spent roughly $500 million on Claude in a single month** because it set no usage caps across thousands of staff. A single engineer on agentic workflows can burn thousands a month; multiply by an org with no governance and you get a nine-figure surprise.

Then the reversal that names the mechanism out loud. **Klarna** automated the work of ~700 agents with OpenAI-powered AI and claimed roughly $40 million in savings, then **reversed course and began rehiring** humans, its CEO conceding, *"We went too far."* The AI gave confident, wrong answers on the hard cases: disputes, fraud, hardship, which in a regulated business is not a quality problem, it's a liability. Klarna is not an outlier; it is the leading edge of what's now called the **AI Boomerang.** By one workforce analysis, roughly **two-thirds of companies that laid off workers for AI are rehiring,** and **one-third spent *more* restaffing than they saved.** Forecasters expect about half of AI-driven layoffs to reverse within a year or two. These companies cut *and* overspent, the worst of both columns.

And the empirical spine under all of it: MIT's 2025 study of AI in business, *"The GenAI Divide,"* found that roughly **95 percent of corporate generative-AI pilots delivered zero financial return,** on the order of $30 to $40 billion invested for nothing. The decisive finding is the one that should reorganize your strategy: the failures were **not about model quality.** They came from **forcing AI into existing workflows with minimal adaptation.** That is not a definition of bad luck. It is the textbook definition of dabbling.

Now hear the law inside the receipts, because it is the opposite of what the headlines say. **This is not evidence that AI doesn't work. It is the Dabbler's grave and the Finisher's moat, itemized.** The Dabbler bolts AI on top of an unchanged organization; token costs explode, ROI lands at zero, and the humans get rehired. AI became a cost *added,* never a cost *replaced.* The Finisher redesigns the workflow so AI *replaces* the cost structure, and on the other side of that redesign the same intelligence that bankrupted the Dabbler comes in cheaper than the labor it retired. Same tool. Opposite outcome. The variable is not the model. **It is whether you rebuilt the org around it or strapped it to the side.**

History has run this exact experiment before, and the result is settled. When factories first installed electric motors in the 1890s and 1900s, they bolted them onto layouts designed for a central steam engine and a spider's web of overhead driveshafts, and for roughly twenty years they saw almost no productivity gain. Economists still teach it as *the* productivity paradox. The gains arrived only when factories were *redesigned around* electricity itself: motors placed at each machine, single-story plants laid out for workflow instead of proximity to a driveshaft. The rhyme is exact, and it is not a metaphor. It is the same mechanism wearing different clothes. **AI bolted onto the old organization is the cost wall. AI built into a redesigned organization is the order-of-magnitude gain.** Dabbling is the steam-era factory with an electric motor chained to the old shaft, wondering why the bill went up and the output didn't.

So when your strategy team brings you a deck full of pilots, ask the one question that separates the 5 percent from the 95: *Has a single one of these redesigned a workflow, or have we merely strapped intelligence to the side of the way we already worked, and started paying for the privilege?*

The Kodak Catastrophe — A Complete Autopsy

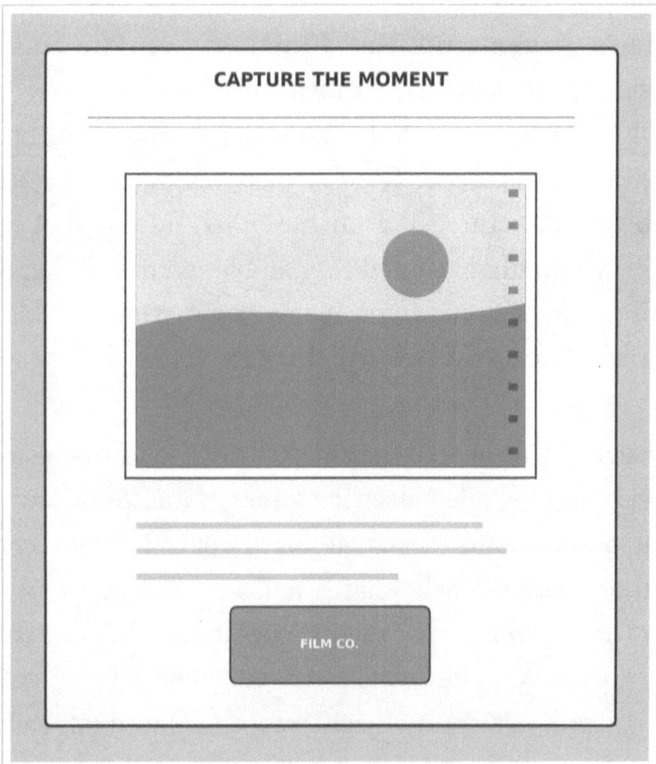

I have referenced Kodak throughout this chapter. Now the full autopsy, because it is the purest specimen of the Success Trap ever recorded: every mechanism, every act of the denial script, every rational decision that turned out to be lethal, all in one body.

At its peak Kodak was an empire. In 1976 it held about ninety percent of U.S. film sales and eighty-five percent of camera sales, eventually employed a hundred and forty-five thousand people, and posted revenue around five and a half billion dollars, roughly thirty billion in today's terms. "Kodak Moment" had entered the language as a feeling, not a product. The moat looked impregnable: a century of chemical expertise, manufacturing scale nobody could match, distribution into every drugstore in the developed world, and a recurring-revenue model that was a work of art. Buy the camera once, buy the film forever. Every birthday, every wedding, every vacation refilled the coffers.

You already watched them bury Sasson's prototype at the top of this chapter, his forecast proven almost to the year. Hold that scene; the autopsy that follows is everything that happened to the body afterward.

Over the next two decades Kodak did everything wrong, and the gut punch is that *everything it did was rational.* In the 1980s it developed digital internally and filed over a thousand digital-imaging patents, then deliberately crippled its own digital cameras to protect film. Imagine being the engineer ordered to make your product *worse* on purpose. In 1988, instead of funding digital, the company decided it was "really a chemical company" and bought Sterling Drug for about five billion dollars: no synergy, vast distraction, eventually sold at a steep loss. Billions that could have bought the future, spent fleeing it. Through the 1990s its argument was that digital would never match film, and it was *technically correct.* Film held more information, richer color, better range, and the professionals agreed. But consumers did not care. They wanted convenience over fidelity, instant viewing over the one-hour photo, free pixels over paid prints. Kodak kept building the world's finest typewriter while the world bought word processors.

Then the collapse, and it was breathtaking. By mid-decade, digital revenue crossed film revenue, and film's collapse steepened from there. The workforce fell from a hundred and forty-five thousand toward seventy thousand. In 2004 Kodak was removed from the Dow Jones Industrial Average after seventy-four years. In 2012 it filed for bankruptcy.

The convenient narrative is that Kodak "failed to innovate." That narrative is not merely wrong; it inverts the lesson and is therefore dangerous. **Kodak invented digital photography.** It held the foundational patents. It employed the best imaging engineers alive. It saw the future *more clearly than anyone.* What killed Kodak was success. Its identity as a film company made it constitutionally unable to become a digital company. Its antibodies strangled every digital initiative that threatened film. Its best customers, the professionals, didn't want digital. Its financial model made protecting film the only rational choice on every spreadsheet. And its leadership ran the denial script with textbook fidelity. Kodak had everything. It invented the future and buried it to protect the present. That is the whole disease, in one company.

Blockbuster, Nokia, and the Near-Miss That Still Missed

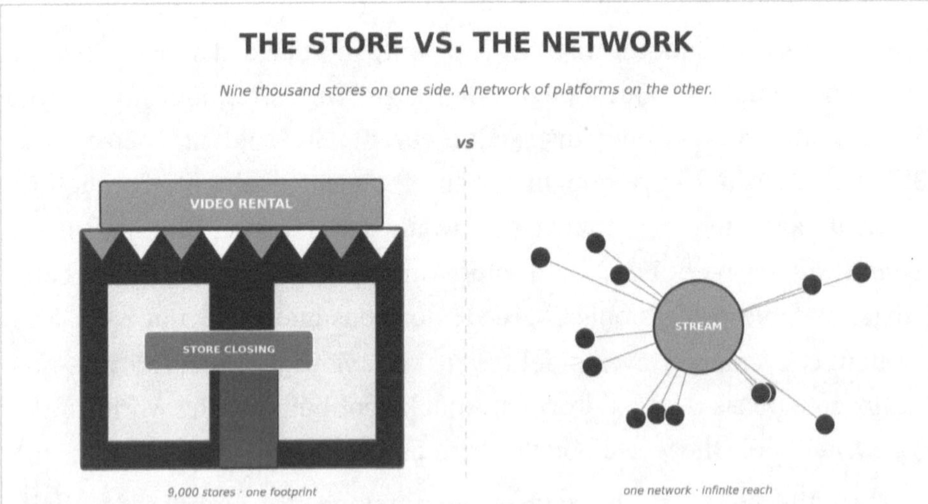

They had nine thousand stores. The future needed none of them.

Blockbuster had, at its peak, about nine thousand stores and sixty thousand employees. Netflix's Reed Hastings walked into its Dallas headquarters in 2000 and proposed a partnership: Netflix runs the online and mail-order side, Blockbuster runs the stores, together they own the market. Blockbuster's CEO reportedly struggled not to laugh. Netflix was small and losing money, the stores were packed, and late fees alone threw off something like eight hundred million dollars a year. When Hastings countered with an offer to sell Netflix outright for about fifty million, Blockbuster declined. Too small. Unproven. Our members love the stores.

Every mechanism fired in sequence. **Identity:** "we are a store company." **Antibodies:** "real estate, store managers, the whole supply chain depend on physical locations." **Customer trap:** "our members love browsing." **Financial gravity:** "eight hundred million in late fees is worth more than any streaming fantasy." **Denial:** "streaming is niche." By 2010 Blockbuster was bankrupt. Netflix today serves more than 325 million paid subscribers and is worth hundreds of billions of dollars.

Nokia's version ran faster. In 2007 it sold forty percent of the world's phones, with the deepest distribution and the strongest brand in mobile. When Apple announced the iPhone, Nokia's internal analysis was *correct on its own terms*. The iPhone was expensive, lacked a keyboard, and made worse calls. But the iPhone wasn't competing on phone metrics; it was laying down a platform. Nokia optimized the wrong variable while Apple and Google built the ecosystem that made the variable irrelevant. Nokia went from forty percent share to near-zero in roughly five years and sold what remained to Microsoft in 2013 for about seven billion dollars. At its peak it had been worth over two hundred billion. Three companies, three industries, three decades apart, the identical script.

And the most dangerous variant of the trap is not denial at all. It is **late recognition.** In the early 1990s, Bill Gates named Microsoft's nightmare competitors as IBM and Novell, rational rivals inside the world Microsoft already ruled. The internet was not the headline threat because the internet was not yet the headline platform. Then the web went from academic curiosity to mass adoption faster than corporate planning cycles could metabolize. By 1994, Microsoft was circulating internal memos that read like a crash course in a phenomenon it had not led; Gates later admitted the internet "mania" had taken him by surprise. The pivot came in 1995 with his "Internet Tidal Wave" memo, framing the internet as the most important development since the IBM PC. And then Microsoft did what great incumbents do when cornered: it fought, hard. It tied Internet Explorer to Windows, used distribution as gravity, squeezed rivals through OEM channels, and won battles.

But platform shifts punish anyone who confuses **distribution control with platform control.** Microsoft treated the internet as a feature to integrate into Windows; the market treated the internet as the operating system of everything, the layer on which search, advertising, mobile, and cloud would compound. The browser wars were won and the war was lost: the browser eventually consolidated around Google's Chrome, and Microsoft ultimately rebuilt Edge on Chromium, the same open-source engine beneath Chrome, a public admission that the platform had moved elsewhere. **The Success Trap is not only "we didn't see it." Sometimes it is: we saw it late, we fought brilliantly, and we still lost the compounding layer.**

Intel is the live case, and its verdict is still being written. Intel dominated the CPU for four decades: the brain inside nearly every PC on Earth, "Intel Inside," fabrication plants that were engineering marvels, legendary margins. When GPUs began winning parallel-computing workloads, Intel's leadership dismissed them as gaming hardware; serious computing meant CPUs, and Intel's entire identity, revenue model, and culture were built on the central processing unit. Meanwhile Jensen Huang made a bet that looked insane: he repositioned NVIDIA's gaming GPUs as general-purpose parallel processors and poured billions into CUDA, the software layer that let developers run scientific computing, and eventually AI, on them. When deep learning exploded in the mid-2010s and every model needed massive parallelism, NVIDIA was the only company with both the hardware and the software ecosystem ready.

Intel watched it happen with more resources, more engineers, and more manufacturing capacity than NVIDIA had. It chose to protect the CPU cash cow, made acquisitions that went nowhere, and shipped AI accelerators that arrived late and underwhelmed. As of mid-2026, NVIDIA — which had become the first company ever to cross **five trillion dollars** in market value in October 2025 — was several times larger than Intel, which after years of damage was still valued in the hundreds of billions rather than the multi-trillion tier. There are now genuine signs of a narrow recovery: under CEO Lip-Bu Tan, Intel beat 2026 earnings expectations sharply, saw its best single trading day in decades on the news that NVIDIA itself was taking a multi-billion-dollar stake, and refocused around inference and agentic-AI silicon. Intel may be the rare incumbent that escapes the trap late rather than never. But "may, barely, after a decade of damage" is the verdict the Success Trap hands its survivors. The lesson is already carved, whatever the epilogue: even a four-decade monopoly with superior resources will protect its cash cow straight into irrelevance, unless leadership chooses, against every instinct, to cannibalize itself first.

The Rare Survivors — and the Five Things They All Did

The Success Trap is a law, not a death sentence. A precious few escape, and they matter precisely because they prove escape is possible while showing exactly how brutal the price is.

Microsoft under Satya Nadella is the cleanest case. In 2014 the company was imprisoned in its Windows-centric identity, the stock flat for a decade, losing platform wars to Google and Apple on every front. Nadella did the thing almost no one does: he changed the identity.

"Windows first" became "cloud first." He embraced the open source Microsoft had once called a cancer, partnered with companies it had tried to destroy, shifted the culture from know-it-all to learn-it-all, and *sacrificed Windows revenue to build Azure*. The results are staggering. Azure became a business worth tens of billions, the stock rose many-fold, the market cap climbed into the multi-trillion-dollar range, crossing three trillion during this era, and the company went from irrelevant in mobile to dominant in cloud to leading in AI inside a single decade. What made it possible: Nadella came from outside the Windows core, Gates was gone from daily operations, there was no line of retreat, and they were willing to cannibalize Windows before anyone else could.

Netflix disrupted itself from DVD-by-mail to streaming while the DVD business was still profitable and growing. Hastings bet the company on streaming even though it meant destroying what worked, and the transition was savage: the stock fell roughly three-quarters, the company shed hundreds of thousands of subscribers, and the press wrote its obituary. Hastings never wavered. Netflix now serves well over three hundred million subscribers.

Adobe's Creative Cloud transition may be the most relevant of all for anyone facing AI. In 2013 Adobe killed the perpetual license, no more buying the roughly $2,500 Creative Suite outright, and went subscription-only at about fifty dollars a month. Customers revolted; tens of thousands signed petitions, the stock dropped, competitors smelled blood. CEO Shantanu Narayen never wavered. Revenue compressed, then grew from roughly four billion dollars to several times that, with recurring revenue climbing above ninety percent. And because Adobe had *already* rebuilt itself around the cloud, it was perfectly positioned when AI arrived, folding generative tools into Creative Cloud instead of scrambling to bolt them on.

The common thread across all three is identical, and it is your checklist: **absolute leadership commitment; complete identity**

transformation; burning the boats with no retreat to the old model; the willingness to endure years of financial pain; and patient capital that measures success in decades, not quarters. Miss any one of the five and the transformation fails. There is no four-out-of-five.

The AI Success Trap — Who Is Caught Right Now

The AI shift is minting a new generation of trap victims, and the list should make a confident reader uncomfortable.

The big strategy firms, the McKinsey/Bain/BCG tier, are caught. Their model depends on expensive human analysts producing insight, and AI now does much of that analysis faster and cheaper. They respond with "AI *assists* our consultants," but the junior-analyst layer that is the base of the entire pyramid is the layer AI hollows out. They cannot cut prices ninety percent to match AI's cost without dismantling the partnership economics that make the firm a firm. They are building the world's finest typewriter.

Law firms sit in the same vise: a business built on billable hours meeting a technology that performs document review, research, and contract analysis at superhuman speed (ask David). "AI makes our lawyers more efficient," they say, but efficiency is the *enemy* of billable hours. When a large share of legal work becomes automatable, the leverage model that has defined the partnership for a century stops adding up.

Google is the most surprising name on the list. Search throws off well over two hundred billion dollars a year in advertising, and AI that answers directly instead of returning ten blue links threatens the entire mechanism: fewer searches, fewer ads. Google is bolting AI overviews onto search, a half-measure that may *accelerate* the very erosion it's meant to slow. It has the best AI talent on Earth, effectively unlimited capital, and the most powerful distribution in technology history. And it may still lose this layer, because the Success Trap does not care about your resources. It cares about your willingness to destroy your own cash cow, and a two-hundred-billion-dollar cash cow is the hardest animal on Earth to put down.

Higher education faces the existential version. The four-year degree is priced on the scarcity of knowledge and expert instruction; AI tutors deliver personalized, adaptive teaching around the clock at near-zero marginal cost. "The campus experience justifies the cost," administrations say. But when the cost structure is tenured faculty, administrative bloat, and physical plant, and the alternative is a tutor in every pocket, the math that justifies the tuition stops closing.

Every one of these institutions is running the exact Kodak playbook: dismiss the technology as inferior, protect the core, listen to today's customers, trust the financial model that says stay the course, and sleepwalk toward irrelevance with improving quarterly margins.

Are You Trapped? — The Diagnostic

The trap is insidious because every instinct it triggers feels like wisdom. The warning signs feel like prudence; the rational arguments feel like strategy; the financial models feel like rigor. Here is how to tell, honestly, whether you are already inside it.

- **If you are dismissing AI** because "the quality isn't there" or "customers don't want it" or "it's hype," you are in the **denial phase.** Every incumbent that ever died said exactly this about the thing that killed it. The telegraph operators said it about the telephone. The gas companies said it about electricity. Kodak said it about digital. The quality always improves, and always faster than you expect.
- **If your strategy is built around protecting the core** ("we can't cannibalize revenue," "the margins are too thin," "we should focus on our strengths"), you are in the **financial gravity well.** The numbers always argue against transformation. They are always right in the short term and fatal in the long one.

- **If your AI program is a pilot, a separate innovation lab, or a "wait and see,"** you are deploying **half-measures,** and you are very likely a Dabbler accumulating the receipt from earlier in this chapter. Half-measures have no historical exception. They fail every time.
- **If you are listening mostly to your current customers and they're telling you they don't need AI,** congratulations, you've found the **customer trap.** The customers who will define the next era are not yet talking to you. They're talking to your future competitor.
- **If you cannot articulate what your company becomes if your core product vanished tomorrow,** you are in the **identity prison.** The most important sentence a leader can complete right now is: *"Who are we if we're not what we've always been?"*

And here is the part nobody says aloud: **the Success Trap does not only catch companies. It catches people.** The lawyer who spent twenty years mastering legal research, who is he when AI does it in seconds? The radiologist who spent two decades reading scans, what happens when the machine reads them more accurately? The accountant whose worth rests on knowing the tax code, what happens when AI knows every regulation in every jurisdiction in real time? The personal trap is harder to escape than the corporate one, because a company can fire its CEO, restructure, and rebrand, but you cannot fire yourself, and you cannot swap out your own identity. When your sense of who you are is fused to a skill being automated, the trap threatens more than your income; it threatens your *self.* Which is exactly why so many capable individuals answer AI the way Kodak answered digital: dismissal, denial, half-measures, and then panic. The escape is the same at the personal scale as the corporate one: stop defending the skill, and start rebuilding around the new platform before it forces your hand.

Why It's Worse This Time

The AI shift is more dangerous than any that came before, for four compounding reasons, and they stack.

Speed is unprecedented. Prior shifts handed companies ten to twenty years to adapt. This one offers three to five. By the time the threat is certain, the window to answer it is nearly shut.

Scale is universal. Previous disruptions hit one industry at a time (manufacturing, then media, then retail, then transport), so companies in untouched sectors could watch from the bleachers. AI hits every industry at once. There is no sideline and no safe harbor.

The economics are brutal. AI can collapse marginal costs toward zero while multiplying speed. These are not incremental gains; they are order-of-magnitude shifts that render existing business models obsolete on a timescale boards are not built to process.

Network effects are decisive. AI platforms reward data, users, and integration with compounding returns: winner-take-all dynamics stronger than anything we've seen. First movers gain advantages that latecomers cannot buy back. There is no comfortable second place.

Stack them: the most powerful disruptive force in history, hitting every industry simultaneously, at a speed that makes deliberate adaptation nearly impossible, with economics that obsolete the old models and network effects that hand early movers an unassailable lead. Your window might be three years.

The Success Trap is optional. Transformation is possible. But only if you are willing to destroy what you've built in order to become what you must, and only if you start now.

In the next chapter we leave history behind and look directly at the engine driving all of this: what AI actually *is*. Not the Terminator fantasy, not the "it's just autocomplete" dismissal, but the verifiable reality of what these systems can do as of mid-2026, and what that means for every industry on Earth. Because the world has split into two camps that are both wrong. One imagines sentient machines next Tuesday and freezes, too afraid of what AI *might* become to act on what it *already is*. The other waves the whole thing off as hype and keeps doing what it always did, certain the bubble will burst, and becomes Kodak, sure the threat is twenty years out right until it's standing on their chest. The Finisher does neither. The Finisher looks with clear eyes and asks the only two questions that matter: *What can it actually do right now? And how fast is it improving?* That is what we answer next. You cannot finish a platform shift you don't understand.

Key Takeaways

- The Success Trap kills market leaders through five mechanisms that operate together with the reliability of gravity: the **Identity Prison**, **Organizational Antibodies**, the **Customer Trap**, the **Financial Gravity Well**, and the **Pattern of Denial.** It is structural, not a leadership failure. Swap every executive and the outcome holds.
- Kodak invented digital photography in 1975 and buried it to protect film margins. It saw the future *more clearly than anyone* and chose the present: the purest specimen of the trap ever recorded.
- The trap is not history. JPMorgan's COIN (reported 2017) did ~360,000 hours of contract work in seconds; the managing partner who runs the math still puts the phone down, because acting means dismantling the machine that made him rich.
- **The Dabbler's Receipt is in:** ~95% of corporate AI pilots return zero (MIT, 2025) because they force AI into unchanged workflows; Uber's annual AI budget was gone by April; Microsoft cut its own engineers' Claude Code over cost; Klarna reversed its AI layoffs and is rehiring. Like steam-era factories strapping electric motors to the old driveshaft, **AI bolted on is the cost wall; AI built into a redesigned org is the 10x.**
- Escape demands all five, with no exceptions: absolute leadership commitment, complete identity transformation, burning the boats, enduring years of financial pain, and patient capital measured in decades. The through-line of Microsoft, Netflix, and Adobe.
- It's worse this time on four stacking axes: speed (3–5 years, not 10–20), universal scale, near-zero-marginal-cost economics, and winner-take-all

network effects. There is no sideline and no comfortable second place.

Finisher Move: Name the single revenue line your organization would never cannibalize, the one that is "too profitable to touch." That is not your fortress. That is the exact coordinate where the disruption lands first. Stand up a small, separate team this quarter whose only mandate is to build the thing that kills it, before someone outside your walls does.

Boardroom Question: *"What is our most profitable product or process that an AI-native competitor could deliver for one-tenth the cost – and have we redesigned a real workflow around AI, or merely bolted it onto the way we already work and started paying the bill?"*

Sources & Currency Note

This book states facts as of mid-2026; specific figures move, the pattern does not. Key data points in this chapter: JPMorgan's COIN reviewing ~12,000 commercial credit agreements / ~360,000 annual lawyer-hours (reported Bloomberg, 2017); NVIDIA as the first company to cross a $5 trillion valuation (October 2025) versus Intel in the low hundreds of billions, with Intel showing a 2026 earnings-led recovery under CEO Lip-Bu Tan; the Dabbler cost-wall cluster – MIT's "The GenAI Divide" (~95% of corporate GenAI pilots return zero, 2025), the Catanzaro/NVIDIA "compute costs more than the employees" quote (as reportedly attributed), Uber's 2026 AI budget exhausted by April, Microsoft cutting internal Claude Code licenses ($500–$2,000/engineer/month, cutoff June 30, 2026), the reportedly ~$500M-in-one-month Claude overspend (by one account), and Klarna's AI-layoff reversal / the broader "AI Boomerang" (per Careerminds: ~2/3 rehiring, ~1/3 spent more than they saved); plus the survivor transformations of Microsoft (Azure; a multi-trillion-dollar cap, having crossed $3T), Netflix (>325M subscribers), and Adobe (perpetual-license kill in 2013, revenue growing several-fold to recurring revenue above 90%). The factory-electrification "productivity paradox" (~20-year lag until plants *were redesigned around electricity) is the historical anchor for the cost-wall thesis. Where a single figure could age badly, it is framed as a range or dated explicitly.*

CHAPTER 4

INTELLIGENCE RISING

> **IN THIS CHAPTER** the one capability that makes this platform shift different from every shift before it — *intelligence itself is being democratized, priced toward zero, and poured into the walls of the economy.* **Why it matters:** every prior revolution made one input abundant; this one makes abundant the input that decided every other contest in history. **What to do with it:** stop measuring AI against the work you do today, and start positioning for the world where expert-grade thinking costs a fraction of a cent and arrives in seconds.

Six platform shifts made one thing cheap. The telephone made distance cheap. Electricity made power cheap. The automobile made movement cheap. Computing made calculation cheap. The internet made communication cheap. Mobile made all of it portable.

The seventh shift is not making a thing cheap. It is making intelligence cheap, the single input that has decided the outcome of every other contest in human history. Which armies won. Which companies dominated. Which children were educated into the professional class and which were not. Intelligence has always been the foundational advantage, and it has always been scarce, expensive, and slow to manufacture. That is the variable now collapsing toward zero.

This is the chapter where the pattern from the Introduction stops being history and becomes your operating environment. Everything we have named, the Durable Sequence, the open window, the toll and the road, runs on one engine. This is the engine.

The Expert and the Machine

A radiologist has been reading chest CTs for twenty years.

She is one of the best in her hospital system, the one residents bring the scans they cannot resolve, the one who catches what others miss. Her eyes have examined over three hundred thousand images across two decades. She can spot a pulmonary embolism in seconds, and tell the difference between a benign nodule and the early whisper of something that will kill a patient in three years.

On a Tuesday morning in late 2025, she reviews a routine screening CT from a sixty-two-year-old former smoker. She runs her process, systematic and thorough, the same way she has done it three hundred thousand times, and marks the scan clean.

The AI flags a four-millimeter nodule in the right lower lobe.

She zooms in. Studies it. Adjusts the contrast. And there it is: a faint shadow tucked behind a rib, half-hidden by normal anatomy. The machine caught it because it does not get tired at three in the afternoon. It does not carry a backlog of forty scans creating time pressure. It does not suffer the pattern-matching fatigue that touches every human reader after hour six of a shift.

She sits back. She is not angry, and she is not threatened. She is calculating: how many four-millimeter nodules she has missed in twenty years because she is human and humans miss things. How many patients she sent home with clean scans who came back eighteen months later with stage-three cancer. The mornings with forty scans queued and not enough hours. She thinks about the years and the cost of the training that made her the expert she is. And she thinks about a system trained in a matter of months that can already see what she cannot.

What matters about that moment is not the part your strategy team will fixate on. She does not report the AI's flag with resentment. She reports it with relief, because for every nodule the machine catches that she missed, a patient gets treated in time. She makes the AI a first-pass filter on every scan. Her accuracy rises. Her speed rises. Her confidence rises. Twenty years of judgment plus tireless pattern recognition catches more than either alone. Not marginally more, materially more.

She is not the victim of this story. She is the model for it. But understand exactly what just happened, because it is happening this quarter in radiology suites, trading floors, law firms, and research labs on every continent. A world-class expert met a machine that performs a core piece of her expertise at superhuman consistency, for a rounding error in cost, and the value of her two decades did not evaporate. It moved. It moved out of the execution she had been paid for and into the judgment only she could sign. That migration is the whole subject of this book, and intelligence becoming cheap is what forces it.

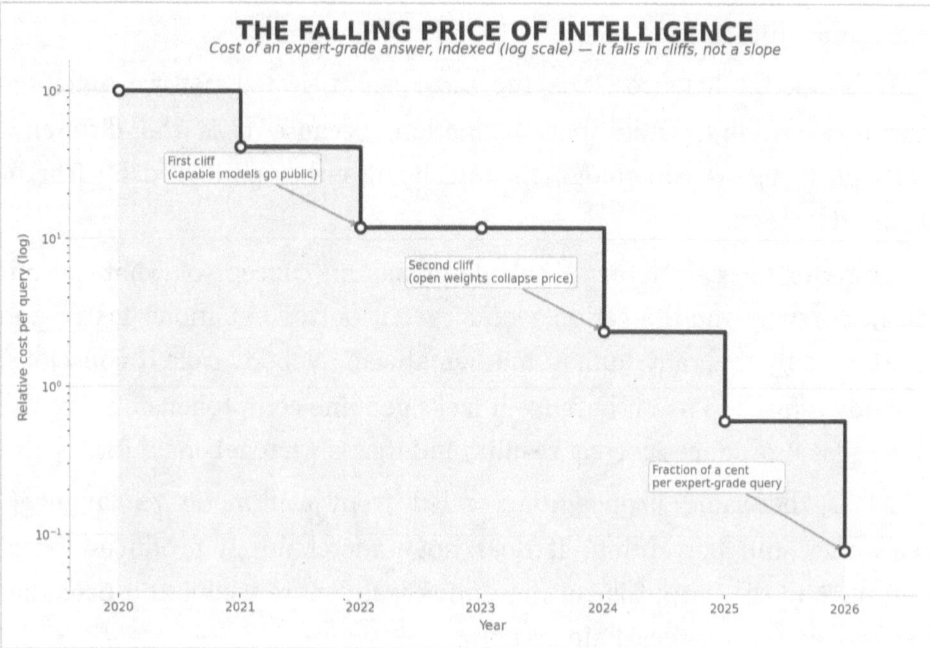

What AI Actually Is — Stripped of the Hype

Before you can position around a force, you have to see it without the fog. And no technology in history has generated more fog than this one. Half the commentary insists sentient machines are about to enslave us. The other half insists AI will cure everything by next quarter. Both are wrong, and both are expensive to believe.

Here is what AI actually is, in plain terms: **pattern matching at a scale no human brain could reach.**

These systems do not understand anything the way you understand things. When the model writes about love, it has never loved. When it explains gravity, it has no sense of what falling feels like. What it does is ingest staggering quantities of human-generated text, images, code, and data, learn the statistical structure underneath, which words follow which words, which concepts attach to which concepts, which patterns recur in which kinds of images, and then generate output that looks remarkably like human intelligence.

It is not intelligence. It is the most sophisticated pattern-matching engine ever built. Hold that distinction, because it is the difference between using AI with judgment and being either paralyzed by fear or seduced by hype.

Think of the calculator. A calculator has no concept of what "seven" means or why multiplication works, yet it solves equations faster and more reliably than any human mathematician. Nobody calls it conscious. Nobody refuses to use it because it lacks genuine comprehension. We use it because it produces correct results, and that is what gets paid for.

AI is the same proposition, scaled from arithmetic to language, reasoning, and perception. It does not understand. It produces useful output. And the output is getting dramatically more useful at a pace that has broken every forecast aimed at it.

Now the architect's correction. The fact that it is "only" pattern matching is not a reason to relax. It is the reason to move. A capability that delivers expert-grade results does not have to be conscious to reprice your industry. The grid did not understand light; it still ended the gas-lighting business. Stop debating whether the machine is "really" thinking. Start mapping which of your processes are, underneath, sophisticated pattern matching, because those are the ones about to be industrialized, whether or not the vendor admits it.

The State of the Frontier

The landscape, as of mid-2026, is a race that keeps accelerating even as the field of runners grows.

A handful of frontier labs, with OpenAI, Anthropic, Google, Meta, and xAI prominent among them, are shipping new flagship models on a cadence measured in weeks, not years. The capability envelope now routinely includes extended multi-step reasoning, agentic coding (models that do not merely write code on request but plan, execute, and debug multi-step software projects), seamless multimodal work across text, image, audio, and video, and context windows large enough to hold the equivalent of several full-length books in a single session. The specific version numbers will be stale by the time you read this. The direction will not be. Plan for the direction.

Two facts cut through the version churn and are worth carving in.

The first is demand. **By May 2026, ChatGPT had more than nine hundred million weekly users and was still approaching a billion** — an adoption curve unmatched at that scale. That is not a product metric. That is more than nine hundred million people, in under four years, rewiring how they get answers. Distribution at that scale is its own moat, and the incumbents with a billion existing users in search, mobile, and social know it, which is why every one of them has bolted a model into the products you already open by reflex.

The second fact is price, and it is the one that should reorganize your strategy. In January 2025, a Chinese lab called **DeepSeek** disclosed a frontier-class model, R1, alongside a widely cited **$5.6 million** training figure — which referred to the reported compute for its V3 base model, not the all-in cost of R1 — at a moment when American labs were spending billions per run. The exact figure was debated and revised in public reporting. What was not debatable was the market's verdict. **The next trading day, January 27, 2025, NVIDIA lost about $589 billion in value — the largest single-day loss by a single**

company in the history of the stock market. That was not a crash. It was intelligence being repriced in public, in a single session.

Read the DeepSeek episode correctly, because most readers drew the wrong lesson. The lesson is not "GPUs stopped mattering." DeepSeek hit its efficiency with clever architecture, activating only the relevant portion of the network for each task rather than firing the whole thing for every query, and a small team using far fewer chips than anyone assumed was required. The lesson is structural: **the cost of intelligence does not decline smoothly. It falls in cliffs.** It drops off a ledge, catches its footing for a few months, then drops off another. The collapse comes from several directions at once: better hardware, better algorithms, better architectures, better training methods, and each advance compounds the others. This is why every linear projection of AI cost has been wrong, and will keep being wrong. If your plan assumes today's prices hold, your plan is already obsolete.

Take three things from the frontier and leave the rest to age out.

First, the capability frontier moves faster than long-range prediction can track. The models of early 2026 make those of early 2024 look primitive; the models of 2028 will do the same to today's. Plan for acceleration, not stability. A roadmap that assumes a fixed capability ceiling is a roadmap to being lapped.

Second, cost is collapsing toward zero, and the collapse is jagged, not gentle. DeepSeek proved even training cost, the most expensive part of the whole enterprise, can be cut by orders of magnitude. The strategist who waits for prices to "settle" is waiting for a thing that does not happen in this phase.

Third, the competition is global, and the strategies are incompatible by design. Some labs are building closed, proprietary systems you rent through an API at their price and under their rules. Others are giving powerful models away as open source, deliberately commoditizing the model layer so no rival can charge a premium for what they hand out free, because if intelligence is free, their core business becomes more valuable. The incumbent search giants are caught in the classic trap: their first concern is not winning AI but protecting the enormous advertising business that AI threatens to undercut. And Chinese labs keep proving frontier work can be done for a fraction of Western spend. American firms, Chinese labs, European startups, and open-source communities are all pushing the boundary simultaneously.

The Durable Sequence tells you how this resolves: the chaotic invention-and-infrastructure phase always consolidates into a small number of platform winners. Two hundred and fifty-plus carmakers became three. Dozens of operating systems became two. A field of search engines became one. It always looks ungovernable from the inside. It always resolves faster than the participants expect. We are not there yet. The battle is live, the strategies diverge violently, and the eventual winner could come from any direction, which is precisely why the positions are still open.

What AI Can Do Right Now

Vague claims about AI being "transformative" do not help you allocate capital. Specifics do.

As of mid-2026, frontier systems perform at or near expert level on a widening list of professional benchmarks: passing the bar and medical-licensing exams in the top deciles, writing code at the level of a competent mid-level developer, producing professional-grade marketing copy, legal drafts, research summaries, and financial analyses. They generate photorealistic images from a sentence, produce video, clone voices with unsettling fidelity, translate across more than a hundred languages, and surface patterns in financial data that human analysts miss. These are not lab demos. They are in production, at scale, touching millions of people every day.

Watch coding most closely, because **what is happening to software is the leading indicator for every knowledge profession.** By GitHub's own figures, AI now writes a large and rising share of new code in projects using its Copilot, on the order of roughly half for active users as of 2025–26, up from essentially zero at the tool's 2022 launch. AI-native editors have changed the act of building: instead of writing code line by line, developers describe what they want and the system assembles it. Tools now let non-programmers stand up functional applications by describing what they need in plain language. The wall between "technical" and "non-technical" is dissolving. A marketing manager who can specify clearly can ship in a weekend what once required a small team and a six-figure budget. That is democratization in its rawest form: a scarce, gated capability becoming available to anyone who can articulate intent.

In customer service, the shift is already structural in many organizations. The Swedish payments company **Klarna** automated the work of roughly **seven hundred agents** with an AI assistant, then reversed course and began rehiring when full automation degraded the hard, emotionally charged cases, the reversal Chapter 3 dissects in full. The durable model it landed on is the one to internalize: **AI for volume, humans for nuance.** Industry-wide, a large share of routine queries now resolves with no human in the loop, and the ripple does not stop at the company line. Entire national economies built on business-process outsourcing are facing a reckoning as the work that employed millions gets automated.

In knowledge retrieval, the old ritual of typing keywords and receiving ten blue links is dying. Answer engines now synthesize across the web and respond directly, and the incumbent search giant has been forced to put AI-generated answers above its own links, cannibalizing the very business model it is trying to defend.

And the current frontier is **agents**: systems that plan and execute multi-step tasks rather than just answering questions. This is the most consequential capability shift since the models first became useful, because it is the line between a tool you operate and a worker you direct. An agent can research a topic, draft the report, route it for review, and book the follow-up, all from a single instruction. It can triage your inbox, flag what matters, draft responses, and run routine correspondence. It can manage a workflow and coordinate across a team.

Today's agents are still bounded. Error rates remain too high for unsupervised use in high-stakes work; when an agent fails partway through a ten-step task, you often do not learn it failed until you inspect the final output. They make mistakes a human never would, not errors of fatigue but errors of misunderstanding. An agent told to "clean up the database" may delete records any employee would instantly recognize as critical, because it lacks the contextual knowledge that comes from years inside a business.

But here is the line that should change how you plan: **the gap between "unreliable agent that needs supervision" and "reliable agent that handles routine work alone" is a software problem, not a physics problem.** Software problems get solved, and they get solved fast, and when one is solved, the improvement is instant and global. No retraining period. No learning curve. No onboarding. One version update, and every agent on the platform is better at once. This is the moment AI crosses from "tool I use" to "colleague I direct," and tools raise your productivity, while colleagues change the structure of the work itself. When agents become reliable enough to run the routine knowledge work that consumes the majority of most professionals' days, the definition of a knowledge worker changes. What remains is judgment, creativity, and relationship: the work that requires a human who can be held responsible. That barbell is arriving faster than any HR department is staffed for, and we map it in the next chapter.

What AI Cannot Do — The Honest Truth

This is where most books on AI lose the plot. They either tell you the machine can do everything (a lie) or that it will never match humans (also a lie). The truth is more precise, and far more useful, because the limitations are exactly where your defensible value lives.

AI has no consciousness. It experiences nothing. When it writes about grief it has never grieved; when it describes chocolate it has never tasted anything. So any work that turns on genuine empathy, such as therapy, counseling, pastoral care, and end-of-life support, requires a human who has actually felt what the person across the table is feeling.

AI hallucinates. It states falsehoods with total confidence, and there is no reliable way to know it is wrong without independently verifying the output. So any task where accuracy is non-negotiable, such as diagnosis, legal advice, and financial reporting, still requires human verification. AI makes those tasks faster. It does not make oversight optional.

AI has no common sense. It lacks the basic model of the physical and social world that a young child carries effortlessly. It can recite that water flows downhill without any grasp of why. A two-year-old who has never seen a glass fall knows it will break; the model must be trained on countless examples to reach the same conclusion, and still cannot reliably generalize to a situation absent from its training data. So it struggles with the genuinely novel, the edge case, the scenario no dataset anticipated.

AI does not learn from your relationship with it. Each interaction stands largely alone; the system that helped you today does not remember yesterday unless you re-supply the context. Context windows are growing dramatically, and a frontier model can now hold the equivalent of several novels at once, but that is a workaround, not the deep, evolving understanding that accumulates between people over years. The physician who has treated you for two decades knows things no record captures: your anxiety patterns, your habit of minimizing symptoms, how stress shows up in your body. That accumulated relational knowledge is not in the model.

And the deepest limit of all: AI cannot originate. It recombines patterns from its training in novel ways, producing variations and juxtapositions no human would have assembled, but it cannot have the conceptual breakthrough that comes from a conscious mind wrestling with a problem: the leap that connects two distant domains, the insight that breaks with everything prior. Special relativity did not emerge from pattern-matching the existing physics literature. It came from a human imagining what it would be to ride a beam of light.

These limits are real, and you should build around them. But here is the error nearly everyone makes: **they treat the limitations as fixed.** They look at what AI cannot do today and assume it never will, the exact mistake the telegraph operator made about the telephone. In 1876 the telephone could not move information as fast as Morse code. By 1900 the telegraph was dying.

AI's limitations are real. They are also shrinking, fast.

Consider hallucination. As recently as 2023, leading models stated confident falsehoods at rates that ranged from a few percent on general knowledge to well over half on specialized professional questions. By mid-2026 the best general-knowledge error rates have fallen into the low single digits, lower still when the model can consult real-time information and verification tools. Low single digits is still too high for unsupervised diagnosis or legal advice. It is more than low enough for the vast majority of knowledge work, and the trajectory is unmistakable.

Context tells the same story. In 2023 most models held a few thousand tokens, a handful of pages. Within two years that stretched into the hundreds of thousands; the frontier now reaches into the millions, several full-length books in one session. An AI that can read your entire annual report, every contract, all your customer correspondence, and every competitor filing at once is no longer constrained by what it can see. It is constrained only by what it can understand, and understanding is improving faster than capacity.

So treat the limitations as a moving front line, not a moat. Each is real today. Each was worse last year. Each will be smaller next year. **A strategy built on AI's present weaknesses is a strategy built on sand**, because the weaknesses are the first thing the next release erodes. Build instead on the limits that are structural rather than temporary, because those are the ground that holds.

> REALITY CHECK AI fails at genuine origination, physical-world judgment, ethical reasoning under novel conditions, and the building of trust. These are not bugs awaiting a patch. They are what intelligence *without consciousness* looks like. The professionals who endure are the ones whose value lives in those gaps. Audit your own work and ask which of your contributions sit in the shrinking zone the next model will absorb, and which sit in the structural zone it cannot.

The Trust Crisis — What Democratized Intelligence Breaks

The cheerleaders leave one thing out, and it bears on your strategy as much as any capability gain. When you democratize the power to generate convincing reality, you democratize it to everyone, including the people lying to you.

Recall the finance clerk from the Introduction: the **$25.6 million** wired on a video call where every face, including the CFO's, was synthetic. He recognized them, questioned them, got fluent answers, and authorized the transfers anyway, because the reality he was evaluating was not real. Twenty years of reading people across conference tables counted for nothing against a technology that fabricates people indistinguishable from the originals.

That incident is no longer exotic. Across 2024 and 2025, executives have been deceived by deepfaked video calls impersonating partners and counterparties; voices have been cloned from public recordings to authorize wire transfers; procurement teams have been targeted by synthetic video of their own CEOs ordering emergency payments. In each case the targets did exactly what diligence demands, verifying the face, recognizing the voice, confirming the context, and in each case it was insufficient. **The attack surface is not human gullibility. It is human trust itself**, the intuitions about identity and presence that have been reliable for the entire history of our species, now exploitable at industrial scale.

Sit with the structural consequence. If a trained professional, in a controlled corporate environment, following established verification procedures, can be fooled by manufactured reality, then **every verification system you currently rely on is already compromised.** Every video call, every voice authentication, every "I recognize that face" is now a potential attack surface. That is not a forecast. It is a present condition.

The damage runs past fraud into the information supply itself. Content floods are overwhelming human capacity to judge quality: a model can produce a thousand grammatical, plausible-looking articles an hour, and the signal-to-noise ratio of the open web is collapsing precisely as the volume goes infinite. Expertise is being diluted, because when anyone can generate expert-sounding content on any topic, distinguishing real competence from confident synthetic output becomes a genuine problem. An AI-written medical article can carry citations, caveats, and clinical reasoning, and stand behind none of it, because the system has no understanding of what it wrote and no accountability for being wrong. The arms race between generation and detection is tilting decisively toward generation: making fake content keeps getting cheaper, catching it keeps getting harder, and that asymmetry will define the information landscape for the next decade.

Now read all of this the way a finisher reads it, because buried in the crisis is one of the largest platform openings of the entire era. When reality can be manufactured, trust stops being a feeling and becomes infrastructure: something that must be built, verified, metered, and owned. Content-provenance systems, authenticated-media standards, cryptographic signatures on every consequential file, verified-human attestation, all of these are being built right now, and whoever builds the layer the rest of the economy is forced to route its trust through captures the value of everything that flows across it. That is not a defensive cost center. That is the iron law of the Durable Sequence pointed at the newest scarce resource: **proof that something is real.**

For your strategy, the trust crisis means three concrete things. **First, your authentic human presence appreciates, not depreciates**, because in a world where anything can be faked, verifiable human connection and accountable human judgment command a premium. **Second, you need verification systems now**, before you are the next deepfake wire or disinformation target. This is a this-quarter procurement decision, not a someday memo. **Third, if you build infrastructure, build the trust layer.** The proof-of-real layer is one of the great unclaimed positions of the AI era, and it is being claimed this cycle.

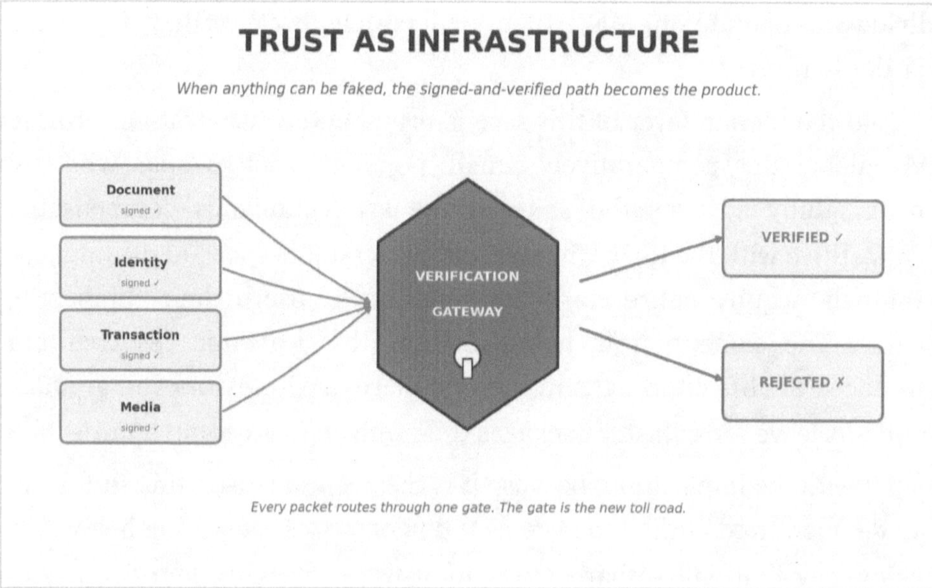

The Geopolitical Dimension

One more reality shapes the ground you are positioning on: this is not only a technology race. It is a geopolitical one, and your AI strategy is now, like it or not, a sovereignty decision.

The United States and China are locked in an AI competition with no clean historical parallel. The U.S. restricts the export of advanced chips; China builds workarounds and pours capital into domestic fabrication; both governments treat AI dominance as a matter of national security, economic competitiveness, and military power, and both are spending on the order of hundreds of billions to win it. The result is a fracturing of the global technology stack into competing blocs. Chinese models operate under different rules, different data policies, and different oversight than American ones; European regulation forms a third distinct regime; large developing markets play both sides. For a business leader this means which models you use, where your data is processed, which rules you comply with, and which ecosystems you build on are no longer purely technical choices. They are decisions about which side of a hardening divide you operate on. Make them deliberately, because they are getting harder to reverse.

And the human layer of this race is brutal and concentrating. Frontier AI talent, the comparatively small population of people who can meaningfully advance the state of the art, commands compensation competitive with the top of finance and professional sport, and companies routinely acquire entire startups primarily to absorb their engineering teams. The same pull is hollowing out the academic research that produced the breakthroughs in the first place, as professors and graduate students leave for industry packages their universities cannot approach.

Sit with the implication, because it is the deepest reason this shift is not yet democratized at the top even as it democratizes everything below. The technology that will reshape every industry, profession, and country on Earth is being designed by a population small enough to fit inside a single concert hall, and their choices about architecture, safety constraints, pricing, and access will decide whether AI becomes the great equalizer or the great concentrator; whether it spreads intelligence outward or hoards it. The stakes have never been higher and the decision-makers have never been fewer. You do not control who sits in that hall. You control whether you build on the rails they are pouring or merely rent them, and that choice is still, for a little while longer, yours to make.

Where AI Sits in the Durable Sequence

Hold the whole map up against the five stages, and the position is unmistakable.

Invention is cleared. Frontier models write, reason, code, and analyze at professional levels; multimodal systems handle vision, speech, and video. The "can it even work" question is closed.

Infrastructure is in full flood. In October 2025, **NVIDIA — which sells the engines this revolution runs on — became the first company ever to cross a five-trillion-dollar valuation.** The largest cloud builders are spending **on the order of $700 billion in 2026 alone** on AI infrastructure, close to double the prior year, and more, each, than most nations spend on their militaries. They are reaching for power wherever they can find it: **Microsoft signed a twenty-year deal to bring Three Mile Island back online – announced in September 2024, the reactor renamed and the restart now targeted for 2027** for no purpose other than to feed the machines. And hundreds of companies are pouring the tooling layer, the APIs, frameworks, and environments that let ordinary developers ship AI applications. Trenches, copper, meters. The toll booths are being set this quarter.

Platformization is just beginning. Cost has fallen by orders of magnitude in a few years and is still falling; more than nine hundred million weekly users by May 2026 signal mass adoption; thousands of AI-native startups are spawning on top of the platforms; the network effects are engaging, where more users make better models and better models draw more users.

If you have ever watched a river approach a waterfall, you know exactly this moment. The current is strong but still navigable. Position yourself now and you ride the rapids. Wait, and you go over the falls. And remember the clock from the Introduction, because it governs everything: electricity took roughly **forty years** from invention to default; computing about **fifteen**; the internet about **ten**; mobile about **five**. AI does not need to wire the streets or lay the fiber, because the prior shifts already did, so it propagates instantly through the billions of screens already in your customers' hands. Its window sits on the compressed end of that curve: **three to five years, total.** The ignition was November 30, 2022. We are roughly **three and a half years in as of mid-2026**, and the decisive phase is the next twenty-four months.

The Economics of Intelligence Approaching Zero

The most important number in AI is not a benchmark score or a parameter count. It is the cost curve, and the cost curve is the engine that drives every other claim in this book.

The price of generating AI output has fallen by orders of magnitude in just a few years, from a figure that made each query a luxury, to one that made it a tool, to one that makes it a default, and it is still dropping in the cliffs-and-ledges pattern DeepSeek made visible. The precise dollar figures are perishable and the vendors revise them monthly. The shape is not perishable, and the shape is the point.

That shape is the oldest story in platform economics. **The cost of the core resource drops exponentially, which triggers an explosion of uses that were not viable at the old price.** When electricity ran from roughly twenty-four cents a kilowatt-hour in 1882 – about eight dollars in today's money – to about a dime by the 1910s, it crossed from a luxury for wealthy homes and factories into the thing every building in the country got wired for. When computing fell from millions of dollars a mainframe to a thousand dollars a PC, it crossed from a corporate asset into a household one. The curve does the work; adoption follows the curve.

Intelligence is now running that same curve, and faster. So apply the architect's reading. At a luxury price, AI was a special project. At a tool's price, a line item. At a default's price, an assumption. At a fraction of a cent, it becomes invisible, embedded in everything, assumed everywhere, like current in the walls. You do not calculate the cost of running your desk lamp. You just use it. AI is heading for exactly that invisibility, and when it arrives, every process that involves information, analysis, or a decision will be augmented by it, not because an executive mandated it, but because the economics will make it insane not to. Just as it became insane not to electrify your factory once power was cheap, even if your steam engine still ran fine.

> SIGNAL TO WATCH When the cost of intelligence approaches zero, every process involving information, analysis, or judgment gets augmented by AI, not on anyone's mandate, but because the math makes refusing it irrational. The moment your competitor's per-decision cost drops below yours and stays there, you are no longer competing on the same cost structure. You are the gas lamp next to the wired street.

This is the **Democratization of Intelligence**, and it is the most consequential economic shift in human history, not a phrase but a mechanism. Every prior platform shift democratized something important: the telegraph and telephone democratized communication, the printing press and the internet democratized knowledge, computing democratized calculation, e-commerce democratized markets. AI is democratizing **intelligence itself**, the foundational advantage that has decided outcomes at every level of civilization, from which child gets educated into the professional class to which company captures a market. And it is becoming as cheap as electricity.

Hold that against the most-cited number in the field. **McKinsey estimated AI could add about $13 trillion to global GDP by 2030.** Take that estimate seriously and then understand why it almost certainly *undercounts*, because it models AI as a productivity tool layered onto today's processes, not as a platform that resets the processes themselves. Every revolution in this book looked modest when it was measured as "the old thing, done faster," and enormous only in hindsight, once the world had been rebuilt around it. The thirteen-trillion figure is the productivity-tool reading. The platform reading is a different order of magnitude, and the finishers who own the rails will capture the overwhelming majority of it.

The question was never whether this shift happens. It is happening. The question is whether you will be a finisher or a footnote.

The radiologist from the opening made her choice. She did not fight the machine, deny what it could do, or pretend her two decades made her immune. She looked at the four-millimeter nodule she had missed and asked the hardest question a professional can ask, *what if the machine is better at this than I am?*, and then the question that separates finishers from victims: *how do I become more valuable by working with it?* Her answer moved her value out of the execution the machine had industrialized and into the judgment only she could sign.

That is the question this whole shift puts to you, and it has a deadline. Not whether AI will touch your work, because it will. Not whether soon, because it is now. The question is how you position so that you are more valuable *because* of cheap intelligence, not less valuable in spite of it. The expert who asks it early owns a road. The one who waits for certainty rents it.

In the next chapter, we make the mechanism concrete: exactly what changes when intelligence becomes as cheap and abundant as electricity, and why that single fact will reshape your industry more profoundly than any technology shift that came before it.

Key Takeaways

- Six platform shifts each made one input cheap; the seventh makes **intelligence itself** cheap, the foundational advantage that has decided every prior contest in history. That is why this shift dwarfs the others.
- AI is sophisticated **pattern matching, not understanding**, and that is precisely why it can reprice your industry without being conscious. Map the parts of your work that are, underneath, pattern matching; those are first to be industrialized.
- The cost of intelligence falls in **cliffs, not slopes**: DeepSeek's disclosure erased ~$589B of NVIDIA's value in a single day (Jan 27, 2025). Any plan that assumes today's prices hold is already obsolete.
- AI's limits, namely origination, physical-world judgment, novel ethical reasoning, and building trust, are **structural where it counts and**

shrinking everywhere else. Build your defensible value on the structural gaps, never on the shrinking ones.
- Democratized generation breaks trust, which makes **proof-of-real its own infrastructure layer**, one of the great unclaimed finisher positions of the era. The window is **~3–5 years, opened Nov 2022, decisive phase through the next ~24 months.**

Finisher Move: Take your single hardest *real* deliverable, not a toy demo, and put a frontier model and an agent on it for one focused hour. Measure exactly what it produces, then mark which parts of your own contribution it absorbed and which it could not touch. That line is your future job description.

Boardroom Question: *"Have we tested what AI can do on our most complex, revenue-bearing deliverable — or are we still evaluating it on the tasks we already consider unimportant?"*

Sources & Currency Note

This book states facts as of mid-2026; specific figures move, the pattern does not. Key data points in this chapter: ChatGPT weekly active users (more than 900M by May 2026 and still approaching ~1B, OpenAI); DeepSeek's R1 disclosure (~$5.6M reported training cost — the V3 base-model compute, not R1's all-in cost — January 2025) and the subsequent ~$589 billion single-day NVIDIA loss (January 27, 2025 — the largest one-day loss by a single company in market history); NVIDIA as the first company to cross a $5 trillion valuation (October 2025); 2026 hyperscaler AI capital expenditure on the order of ~$700 billion; the Microsoft–Constellation agreement to restart Three Mile Island Unit 1 (announced Sept 2024, restart targeted for 2028, later accelerated toward 2027); GitHub Copilot's share of AI-written code (roughly half for active users, GitHub's own figures, 2025–26 — drawn from company announcements rather than controlled studies); Klarna's ~700-role customer-service automation and its 2025 reversal toward rehiring; the historical electricity price decline (~$0.24/kWh nominal in 1882, *roughly $8 in today's dollars, to ~$0.10/kWh nominal by the 1910s); McKinsey's ~$13 trillion AI-GDP estimate by 2030 (2018 MGI), which this chapter argues undercounts a platform shift; the Arup deepfake fraud (~$25.6M, reported early 2024); and the compression curve (electricity ~40y, computing ~15y, internet ~10y, mobile ~5y, AI 3–5y; window opened November 30, 2022). Model version numbers move weekly and were deliberately generalized. Where a single figure could age badly, it is framed as a range, an order of magnitude, or dated explicitly.*

CHAPTER 5

THE DEMOCRATIZATION OF INTELLIGENCE

> IN THIS CHAPTER the moment the scarcest resource in human history — intelligence itself — stops being scarce, and what collapses the instant it does. **Why it matters:** every business model you have ever seen, including yours, was secretly a bet that intelligence stays expensive — and that bet is being called. **What to do with it:** locate your profession on the four-stage commoditization clock, then build for the stage *after* the one you're in — because pricing power dies one stage before anyone admits it.

Start with a fact that should rearrange your week, not just your year.

For two hundred years, every fortune, every firm, every professional license, and every org chart was built on one quiet assumption. Intelligence is the scarcest input in the economy, and you have to pay dearly to rent it. That assumption was so reliable you never had to say it out loud. It was the bedrock under the bedrock. A consultant billed by the hour because trained judgment was rare. A law firm leveraged junior associates because the analysis they produced couldn't be had any cheaper. An entire civilization priced its work off the cost of a good human brain.

That bedrock just turned to liquid. And almost everyone with a stake in it is studying the wrong question, "how do I use these new tools?", while the ground itself reprices underneath their feet.

One question actually decides your decade: **when intelligence becomes abundant, what in your business was secretly priced on its scarcity, and what is that thing worth at zero?**

Answer that honestly and you will see your own industry the way a finisher sees it. Not as a thing AI will *improve*, but as a thing AI will *reprice*, line by line, until only the items that were never about cheap intelligence are left standing. Those surviving items are your real business. Everything else was rent you were collecting on a scarcity that no longer exists.

Let me show you the shape of the collapse, because it has a shape, and the shape is the whole strategy.

You Have Seen This Movie – You Just Never Watched the Ending

Every platform shift runs the same democratization arc, and you already know it by heart. A capability that belonged to an elite becomes available to anyone with the platform. Skill that took years to acquire becomes a sentence you type. A price that gated a profession falls toward zero. The pattern is so consistent it is closer to a physical law than a trend, and the only people it surprises are the ones who profited from the version that came before.

Reach back far enough and it gets uncomfortable, because the first knowledge workers in history were destroyed by exactly this move. Around 3000 BC, the power to record and transmit knowledge belonged to a tiny priestly class: the scribes. In Sumer, in Egypt, in every early civilization, scribes held enormous power for one reason. They controlled the technology of writing. Kings depended on them. Temples depended on them. To record a transaction, preserve a law, or send word to a distant province, you needed a scribe at your elbow. Their scarcity *was* their salary. Then literacy diffused, to merchants, to craftsmen, then to everyone, and the scribes' chokehold on civilization broke. Knowledge exploded. And every single time the gate has opened since, the gatekeepers have made the identical argument: *ordinary people can't be trusted with this.* The scribes said commoners would corrupt the records. The telegraph operators said untrained voices would clog the lines. The publishers said amateurs would pollute the discourse. They were always partly right about the mess and always catastrophically wrong about the net result.

Then the modern reruns, each one larger than the last:

- **The printing press democratized knowledge.** The cost of a book collapsed by orders of magnitude over the decades after Gutenberg, and the Scientific Revolution, the Reformation, and the Enlightenment followed, because knowledge was no longer something you had to be admitted to.
- **The telephone democratized communication.** Bell shattered the telegraph operator's monopoly and let anyone who could talk reach anyone else, instantly, at near-zero cost.
- **The internet democratized information.** Before the web, reaching an audience required a printing press, a broadcast license, or a distribution network, and a handful of gatekeepers decided what the public got to see. After the web, anyone could publish anything to the planet for the price of nothing.
- **The smartphone democratized computing.** Before 2007, a computer was furniture. After the iPhone, a supercomputer lived in every pocket, and banking, retail, transportation, and entertainment were rebuilt on the assumption that everyone carried one.

Each shift took something expensive, scarce, and controlled by specialists and made it cheap, abundant, and available to anyone. Each time, the incumbents warned that democratization would bring chaos. Each time they were right about the chaos. And the chaos produced more value than the orderly, gated system ever had.

Now it is happening again. The resource being democratized this time is not knowledge, not communication, not information, not computing power.

It is **intelligence**, the thing that produces all of those. The input to every other input. Which is why this turn is not the next entry in the list. It is the one the whole list was building toward.

What "Free" Actually Detonates

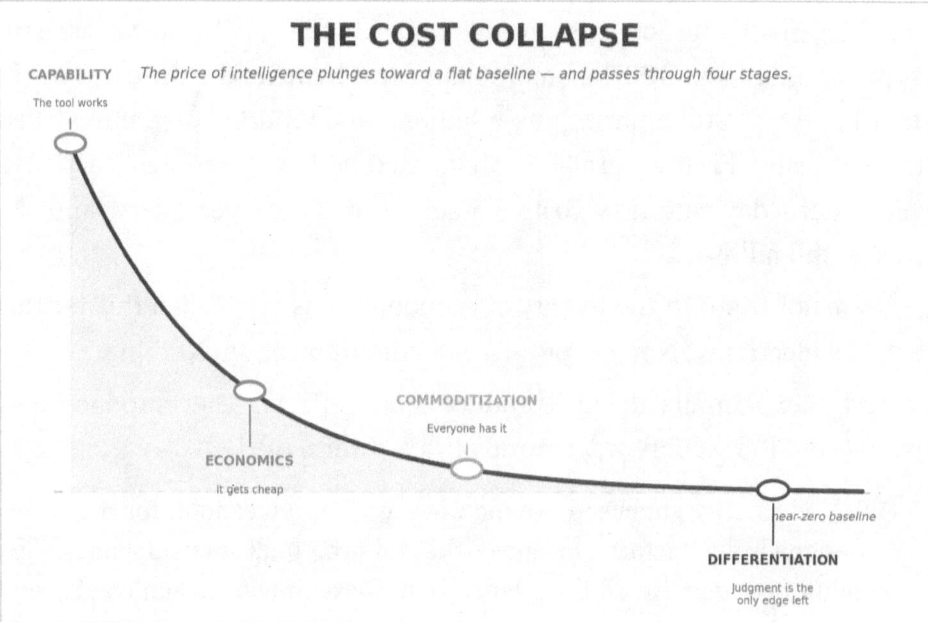

Strip intelligence down to what it is in economic terms: the ability to take in a situation, find the pattern, generate options, and choose well. Every organization on Earth, every company, government, school, and hospital, runs on exactly that. It has always been the scarcest and most expensive input in the economy, for a reason that felt permanent. It required a human brain, and human brains are slow to train, limited in number, unavailable at 3 a.m., and prone to fatigue, bias, and error. The entire architecture of the modern world, education, professional services, the corporate pyramid itself, is scaffolding built around that one scarcity.

So do not ask what happens when intelligence gets *better*. Ask what happens when it gets cheap, as cheap as electricity from a wall.

The cost collapse is the part the mind refuses to hold, so anchor it to something verified. You already watched the clearest proof in Chapter 4: the day a cheap frontier model erased **$589 billion** of NVIDIA's value in a single session, the largest one-day loss by a single company in the history of the stock market. Read that correctly. It was not a panic about one company. It was the market repricing intelligence itself toward zero, in public, in an afternoon. On the demand side the curve is just as violent. By May 2026, **ChatGPT** had more than **nine hundred million weekly users** and was still approaching a billion, an adoption curve unmatched at that scale. Frontier-grade analysis that a few years ago carried a consultant's day rate now costs a fraction of a cent per query, and the price is still falling.

No major input in the history of economics has ever fallen this far this fast. Not electricity. Not computing. Not communication. Nothing.

And cheap inputs do not produce small effects. They produce new worlds, and the second-order world always dwarfs the first.

- When electricity got cheap, we did not merely swap candles for bulbs. We redesigned the factory, invented the suburb, built the appliance, air-conditioned regions of the planet that were previously unlivable, and generated trillions in activity nobody had forecast.
- When computing got cheap, we did not merely retire the typewriter. We built the internet, e-commerce, social platforms, mobile, and the cloud — trillions of dollars of value that no one in 1980 could have named.

AI follows the same law, on a larger scale, for one structural reason. **Intelligence is more fundamental than electricity or computing, because it is the input to both.** When the most fundamental input goes abundant, every output downstream of it improves, not by percentages, but by multiples. That is why the right reaction to "intelligence is getting cheap" is not *"good, my costs drop."* It is *"every price in my industry that was really a price on intelligence is about to be renegotiated without my permission."*

Now, the shape.

The Four-Stage Clock (and Why Pricing Power Dies Early)

Disruption from AI is not random and it is not chaotic. It runs a predictable sequence through every profession, every industry, every skill category, and knowing which stage you are standing in is the difference between getting repriced and doing the repricing. Call it the four-stage clock. **The cruel part is that your pricing power collapses one full stage before the layoffs do, so by the time the disruption is visible, the margin is already gone.**

Stage 1 — Capability. The machine can technically do the task, draft the brief, write the code, build the deck, but the quality is a coin flip. Impressive as a party trick, unreliable as a professional. The whole profession dismisses it on the merits, and the merits are real: *it makes things up, it misses nuance, it's not as good as a person.* All true. And all a trap, because "worse than a human today" gets silently filed as "worse than a human forever." It is not. This was roughly the 2020–2023 window.

Stage 2 — Economics. Quality climbs to something like the *average* human's, and now the spreadsheet takes over the argument. You do not have to believe AI is good. You have to do the math: if it delivers most of the quality at a small fraction of the cost, the pressure is overwhelming and indifferent to your opinion of it. Early movers open a real lead, skeptics start to sweat, and the first wave of quiet displacement begins. Roughly 2023–2025.

Stage 3 — Commoditization. This is the live wire, where much of the professional economy sits as of mid-2026. AI matches or beats the *average* practitioner, and that one word, *average*, is the whole detonation: most professional services are delivered by average professionals. When the machine clears that bar at essentially zero cost, the pricing structure of entire service industries starts to buckle, and clients ask the only reasonable question. Why am I paying three hundred dollars an hour for work a machine does better for three cents? The people who survive this stage are not the ones who *use* AI. They are the ones working visibly **above** the average AI can hit.

Stage 4 — Differentiation. Having capable AI stops being an advantage and becomes table stakes, assumed, like electricity or a network connection. A new hierarchy forms, and it is not built on who *has* AI but on who deploys it with the most judgment, the most strategy, the most unmistakably human edge. The lawyer who survives is not the one who writes the cleanest brief; it is the one whose courtroom instinct, client trust, and strategic nerve cannot be prompted into existence. We are beginning to glimpse this stage on the 2027-and-beyond horizon.

Every profession you can name is somewhere on this clock. Some are still in Stage 2. Many have hit Stage 3. None escape Stage 4. So stop asking *whether* your field will be disrupted. That question is closed. Ask which stage it is in right now, and then commit a heresy most leaders won't: **build for the stage after the one you're in.** If you are priced like Stage 2, your margin is already living on borrowed time, and the move is to abandon the commoditizing middle *before* it commoditizes you, voluntarily, while you still have the cash flow to fund the jump.

> **Sort your own P&L, honestly.** Run a line through your revenue: which of it is "cheap intelligence sold at scarcity prices"? That is the portion the clock is coming for, not someday, but on a schedule you do not control. Everything else is your actual moat. Most leaders cannot tell you the ratio. The ones who can are already repricing themselves on purpose.

The $13 Trillion Number Is Not Wrong — It's Aimed at the Wrong Economy

You have heard that **McKinsey estimates AI could add about $13 trillion to global GDP by 2030.** It has been repeated into conventional wisdom. And it badly understates what is happening, not because the arithmetic is sloppy, but because it is measuring the wrong economy.

The method behind a number like that treats AI as a productivity tool: take today's categories and add a percentage. Ten percent off customer service, fifteen off the supply chain, twenty off drug discovery. Sum the savings. But you read the law in Chapter 1. **Platform shifts do not add percentages to existing categories; they create categories that did not exist.** The automobile did not make the horse ten percent faster; it created suburbs, highways, the oil economy, consumer credit, the shopping mall, and the teenager. The internet did not make the newspaper ten percent leaner; it created Google, Amazon, the streaming business, and the gig economy, every company built on the premise that information is free and abundant.

AI will not make existing businesses ten percent more productive. It will create whole new species of business, new patterns of value creation, new economic structures we cannot forecast, precisely because they depend on the abundance of something that has always, until now, been scarce. What does an economy look like when every small business has the analytical firepower of a top consultancy? When every patient carries diagnostic reasoning rivaling a specialist? When every student has a tutor that adapts in real time? When any founder can build and test a product without writing a line of code?

Nobody can total it, and that is the entire point. You cannot price cheap, abundant intelligence by applying percentages to today's economy, because cheap intelligence will *build tomorrow's* economy, and tomorrow's will not resemble today's enough to be measured against it. The defensible way to say it: $13 trillion is the floor you get from optimizing the world we already have. The real figure belongs to a world that does not exist yet, and it is denominated in categories no spreadsheet has a row for. Do not plan to capture a slice of the optimization. Plan to author one of the categories.

Five Domains, One Verdict

Make it concrete. Here is the democratization already running through five professional domains. Watch the *same* verdict land in each, because once you see the pattern abstractly you will see it instantly in your own field.

Coding. A few years ago there were on the order of twenty-some million professional software developers worldwide, with senior U.S. salaries well into six figures, and the wall between "people who can build software" and "people who can't" was one of the most consequential economic divides on the planet. That wall is dissolving. You can describe an application in plain language and watch the machine build it; among active users of GitHub's Copilot, **AI now writes on the order of 46% of the code** (as of 2025), and purpose-built autonomous coding agents increasingly take a specification, write the application, generate tests, debug their own failures, and ship, without a human touching a keyboard. None of this eliminates developers. It eliminates developers doing *mechanical* work. The skill migrates from implementation to orchestration: the strong developer now directs agents, reviews output, owns the architecture, and spends human attention on the genuinely hard problems, the way a chief executive does not personally write every email. And the total volume of software is exploding, because the startup that once needed five engineers and six months can be one person and a weekend. This is the largest expansion of software creation in history, and most of it is being built by people who would never have called themselves developers.

Writing. Clear, persuasive prose used to command premium prices because it was uncommon. A good post ran tens to hundreds of dollars, a white paper into the thousands. Now anyone can produce grammatical, well-structured, reasonably engaging copy in seconds. The *floor* has risen so fast that what passed for professional-grade writing in 2022 is now indistinguishable from what a free model emits in thirty seconds. The *ceiling* has not moved at all. The best human writers still produce work the machine cannot, work with voice and rhythm and the unmistakable texture of a mind that has actually suffered and celebrated and lived. "Good enough" went free, which means competent writing is now worth nothing and only the exceptional commands a premium. The writer who survives is not the one who can assemble a clean sentence. It is the one whose perspective cannot be regenerated.

Design. Hiring a designer used to mean a hundred dollars an hour and up; a logo ran into the thousands. Now a sentence into an image model returns professional-grade options in seconds for a flat monthly fee, and for routine work, social graphics, decks, product mockups, marketing banners, clients genuinely cannot tell machine from human. Stock photography is dying, because why license a generic photo when the machine renders exactly your scene in exactly your brand's palette? But brand strategy, art direction, and the deep read of human psychology that separates great design from merely competent design are worth *more* than ever. The designer who thrives is not the one with the sharpest tool skills. It is the one who knows why a palette earns trust, why a layout pulls the eye toward the action, why one identity resonates while another dies on contact.

Analysis. Turning raw data into a decision used to require scarce, expensive skill, statistics, business context, visualization, communication, and a six-figure salary to match. Now you upload a spreadsheet, ask in plain English, and get instant analysis with visualizations that once cost a skilled analyst days. The junior-analyst role, standard reports, basic dashboards, routine statistics, is being automated outright, and a single analyst with AI now out-produces the five-person team that came before. But knowing *which* analysis matters, what the numbers mean in context, when the data is quietly lying, and how to decide well on imperfect information, those rose in value. The analyst who survives is the one who can look at a flawless-looking AI output and say: *this is technically correct and strategically meaningless*.

Research. Research once meant hours in libraries, hunting sources, reading stacks, synthesizing by hand, and even after search engines, good research stayed a slow craft of knowing whom to trust and how to find signal in noise. Now research tools deliver instant, cited synthesis across dozens of documents, surfacing findings, flagging contradictions, producing in minutes what took a graduate student weeks. The venture investor who burned forty hours on diligence gets a deeper read in an afternoon. But knowing what to ask, recognizing which findings actually move the decision, and telling important from merely interesting remain stubbornly human. The researcher who thrives is not the one who finds information fastest. It is the one who knows which information changes the decision.

The verdict is identical in all five, and it is the verdict in *yours*. The mechanical skill commoditizes, the baseline rises until everyone is competent, and value migrates from execution to judgment, from being able to do the work to knowing which work is worth doing. **Stop investing in your ability to do the task. Start investing in your authority to decide which tasks are worth doing. That is the only line item the clock does not come for.**

> DECISION POINT Every profession sits on the four-stage clock — Capability, Economics, Commoditization, Differentiation. Name your stage. Then name the *next* one, and ask the only question that matters: what would I have to build this year to be priced for Stage 4 while my competitors are still defending their Stage 2 rates?

The Barbell — and the Trap of the Comfortable Middle

The democratization of intelligence is forging what economists call a barbell economy: both ends grow while the middle hollows out. This is not a forecast you get to debate. It is the mechanical consequence of the four-stage clock running everywhere at once.

At the **high end**, work that demands uniquely human capability becomes *more* valuable precisely because it is now rare against a backdrop of infinite cheap competence: strategic judgment, creative vision, relationship leadership, moral authority, complex physical skill. The executive who can steer an organization through transformation. The clinician whose presence is itself the treatment. The surgeon whose hands do what no machine can. The artist whose vision can't be reached by a prompt. These do not merely survive the shift. They are *bid up* by it.

At the **low end**, AI absorbs routine cognitive work faster, cheaper, and more consistently than people, basic analysis, standard content, rule-based decisions, retrieval, document processing, for fractions of a cent.

And the **middle**, routine professional work that takes real skill but not exceptional judgment, gets hollowed out. Read that sentence slowly, because the middle is where most knowledge workers currently sit, and the middle *feels* the safest. The junior analyst running standard reports. The associate doing document review. The copywriter producing competent, forgettable content. The rep working from a script. These are not the most junior or the least educated roles. They are the *comfortable* ones, well-paid, credentialed, busy, and they are the most exposed in the entire economy. The trap of the middle is that it does not feel like a trap until the clock has already struck.

So here is the uncomfortable truth and the strategy hiding inside it: **when intelligence becomes abundant, human intelligence becomes a luxury good.** Wealthy families will retain human tutors while everyone else gets an AI one. High-stakes decisions will route to human advisors while routine ones route to algorithms. Hand-made judgment will command a premium the way hand-made furniture does in a world of mass production. If you sit in the middle, you do not wait to find out which way it breaks. You pick an end, and you move toward it on purpose, this year, while you still set your own price.

And do not lose the larger truth to the disruption: the *net* effect of all this is overwhelmingly positive, because the total intelligence available to humanity is rising by orders of magnitude. The farmer who could never afford counsel now has legal help in his pocket. The student with no advanced classes now has a tutor at the level of the best private schools. The owner who could never hire a top consultancy now gets strategy for pennies. Democratized intelligence does not promise everyone the same quality. It delivers everyone *access*. And access, historically, is the thing that has transformed civilizations.

Which is exactly why a teenager in a one-room house in rural Rajasthan, with no computer-science classes within two hundred kilometers, no bootcamp, no mentor, and no incubator, but with a smartphone, can describe an idea in her own language, let the machine write the code while she makes every decision that matters, and ship a working product that reorganizes her regional economy. A few years ago that sentence was science fiction. As of mid-2026 it is a category of thing that happens, in places that have never produced a single technology company. *That* is what "intelligence is democratized" means as a life rather than a concept, and it is the supply of new finishers your incumbency now has to compete against.

The Cost of Abundance: When Anyone Can Create, Trust Becomes Infrastructure

There is a darker consequence, and a finisher confronts it directly, because pretending it away does not make it leave. **When the power to create becomes universal, abundance destroys trust, and that turns trust from a feeling you assume into infrastructure you have to build, verify, meter, and own.** This is not a footnote to the democratization story. It is the largest new market the democratization story creates.

Walk the three fronts where trust is already failing:

Synthetic reality. Voice cloning now needs as little as a few seconds of clean audio to produce a convincing replica of a specific person, less than a voicemail greeting. The Arup deepfake from the Introduction, **$25.6 million** wired on a video call where every colleague, the CFO included, was a real-time fabrication, was the first widely-reported taste of it, and the money was never recovered. When you cannot trust your eyes, your ears, or the words on your screen, something load-bearing in how a society functions quietly gives way.

The noise catastrophe. As of 2025, roughly **500 hours of video were uploaded to YouTube every minute**, already more than any human could watch in many lifetimes. Now drop the marginal cost of generating high-quality video toward zero and the flood multiplies. Signal-to-noise does not merely decline; it inverts. Finding genuine, trustworthy information starts to feel like finding one specific grain of sand on a beach. The old internet had a noise problem. The AI-era internet has a noise catastrophe, and "I can't find the real thing anymore" becomes a daily condition rather than an occasional frustration.

The dilution of expertise. When anyone can generate expert-*sounding* work, how do you find the actual expert? A medical post written by an AI-assisted layperson can read like a board-certified physician wrote it. A legal analysis from someone who never opened a law book can read like a partner's memo. Credentials get noisy. The old human tendency to overestimate one's own competence detonates when the machine hands everyone the surface appearance of mastery with none of the underlying understanding, and the person who prompted a sophisticated-looking financial analysis often has no idea whether it is correct, only that it *looks* correct to everyone else.

Society is already groping toward responses, and every one of them is a business: content authentication and provenance standards, proof-of-human credentials, watermarking that separates synthetic from human-made. A flight to quality, when random content can't be trusted, people gravitate back to sources that earned trust over years, the named institution, the handful of voices in any field with a track record. And trust networks, people leaning harder on personal relationships and direct word-of-mouth to decide what's real and who's genuinely competent. In a world where anyone can fake expertise, the people who actually know you become your most valuable filter on the truth.

Now read all three the way a finisher reads them. The lesson is not *"deepfakes are scary."* The lesson is the iron law from the Introduction, pointed at a new surface: **whoever builds the infrastructure that verifies what's real — and that everything else is forced to route through — captures the value of everything else.** The trust crisis is not only a danger to survive. It is the single clearest piece of open infrastructure in the entire shift, and most of it is unbuilt. The trust crisis will not be solved by technology alone; it will be solved the way trust problems always are, by relationships, by institutional credibility compounded over time, and by the slow, unglamorous work of proving yourself through consistent, honest action. Which means the durable position is not a clever detector. It is *being*, or *owning*, the verified source everyone else has to defer to.

The Human Moat — and the Skill That Decides Who Captures the Value

So where does this leave you? If AI is democratizing intelligence, what stays irreducibly human, and what do you actually *do* with it?

Seven things resist the machine. Hold them not as comforts but as the raw material of your moat:

- **Consciousness and subjective experience.** The machine has never felt joy, grief, love, or fear. Any work that turns on genuine empathy requires a human who has actually lived.
- **Moral responsibility.** When a decision carries life-or-death weight, a human has to bear it. You cannot sue an algorithm or hold a neural network accountable. *Someone* must sign.
- **Authentic connection.** As synthetic content floods the world, genuine human presence gets rarer and dearer. The therapist, the mentor, the coach are bid up precisely as the fakes multiply.
- **Physical presence and embodiment.** AI lives in servers; humans live in bodies. Surgery, physical therapy, craft, and athletic performance demand an embodiment robotics is far from matching.
- **Creative origination.** The machine generates endless variation on patterns it has seen; a human can conceive of something genuinely new. Vision, taste, intentionality, and the willingness to risk – the things that make work meaningful rather than merely competent – remain ours.
- **Contextual judgment.** Reading a room, sensing unspoken dynamics, knowing when the standard rule should be broken, navigating real ambiguity: holistic understanding that pattern-matching cannot fake.
- **Meaning and purpose.** The machine processes information; humans make meaning. *Why does this matter? What does it mean for our lives? Where should we go?* Those require a conscious being wrestling with existence. AI can help you process the data. It cannot tell you what the data means for your life.

Now the strategy, stated as law: **do not compete with AI.** You will lose any contest on AI's terms, speed, memory, consistency, scale, availability, the way you'd lose to a calculator at arithmetic, and the loss was never the point. Your move is to become a **centaur**: a human fused with AI, your irreplaceable judgment riding the machine's superhuman processing.

Chess proved this first, and the proof is more profound than the win. In 1997, IBM's Deep Blue beat world champion Garry Kasparov, and everyone declared human chess finished. Instead a new game emerged, freestyle, or "centaur," chess, where human-AI teams faced each other, and the result reorganized everything we thought we knew. The strongest centaur teams beat the best AIs playing alone *and* the best humans playing alone. The combination was simply superior: the human supplied strategic judgment, the sense to override the machine where experience outweighed calculation, and the machine supplied tactical precision the human could never match.

But here is the finding Kasparov himself calls the most significant of the whole experiment, and it is the line you should underline. The strongest teams were **not** the ones with the best human players or the most powerful computers. In a now-famous result, a pair of *amateurs* running ordinary machines beat both grandmasters on supercomputers and grandmasters playing alone. They did not win on chess talent or processing power. They won because their **process for collaborating with the machines was better.** They had built a superior method for consulting multiple analyses, weighing them against each other, and applying human judgment at the exact moments the machines disagreed. The decisive skill was not chess skill or computer skill. It was **integration skill**, the ability to merge human and machine into something stronger than either.

That is the model for every profession in this era. The best doctor will not be the one with the most knowledge or the best diagnostic AI; it will be the one who collaborates best with the machine, who knows when its pattern recognition is trustworthy and when human intuition must override it. Same for the best lawyer, designer, analyst, leader. So this is your operating model, stated plainly: hand the machine everything it does better than you, research, analysis, first drafts, processing, pattern recognition at scale. Spend your human energy on the things only you can do, judgment, relationships, creativity, meaning, leadership. And deliberately build the integration skill itself: knowing what to ask for, evaluating output critically, fusing machine analysis with human intuition. The finishers of the AI era will not be the ones who resist the technology, and not the ones who surrender to it. They will be the ones who learn to work with it so well that they become something neither human nor machine could be alone. And in a world of centaurs, the prize goes not to the smartest human or the most experienced one, but to the best **collaborator**.

Which surfaces the final dimension, the one that decides who actually *captures* the money democratized intelligence creates.

The old digital divide was about **access**: who had a device, who could get online at all. That divide is closing faster than almost anyone predicted. With ChatGPT above nine hundred million weekly users by May 2026 and still approaching a billion and smartphones near-ubiquitous, the access problem, while not solved, is a fraction of what it was a decade ago. But a new divide is opening that is far more consequential, and it does not split on who *has* AI, nearly everyone will. It splits on **who uses it well**, and it is the same brutal 90/9/1 that governs every platform shift in this book, seen now through the lens of who *builds* on AI versus who merely *consumes* it. The Dismissers, Dabblers, and Finishers of the AI tool itself look like this:

- **The ~90% are users**, the Dismissers of this divide. They ask the occasional question, have it draft an email, explain a concept. Real value, modest value, a supercomputer enlisted to check the weather.
- **The ~9% are operators**, the Dabblers. They have woven AI deep into daily work, chaining tools, automating the repetitive, running at several times the output of their peers. They do not just use the machine; they *fly* it.
- **The ~1% are architects**, the Finishers. They are not using AI tools; they are building, combining, and architecting systems that did not exist before. They see AI not as an assistant but as a platform to construct something new on.

That 1% will capture the overwhelming majority of the economic value AI creates, exactly as the 1% who understood the internet as a *platform* in 1995, rather than a faster way to send memos, captured a wildly disproportionate share of its value. The pattern does not change for AI. So the question for you was never whether you'll have access. You will. The question is which tier you operate at, and whether you'll do the deliberate work to move from **user to operator to architect.** That climb is open to anyone willing to make it, and the returns at the top are not incremental. They are the whole prize.

The candlemaker's choice in the Introduction was never about candles. The scribe's was never about clay tablets. And yours was never about AI. It is about whether, when the scarcest resource in history goes abundant, you are still selling the scarcity, or building the thing that's valuable precisely *because* the scarcity is gone.

Turn the page. We pour the foundation next.

Key Takeaways

- **Intelligence is being democratized the way knowledge, communication, and computing were — but this turn is the largest, because intelligence is the input to all the others.** Each prior democratization was bigger than the last; this one is the one they were building toward.
- **Every business model was secretly a bet that intelligence stays expensive.** When it goes abundant, the value migrates from *possessing* intelligence to *deploying* it, and every price that was really a price on cheap intelligence gets renegotiated without your consent.
- **The four-stage clock — Capability → Economics → Commoditization → Differentiation — runs through every profession, and pricing power dies one full stage before the layoffs.** Build for the stage *after* the one you're in.
- **The economy is splitting into a barbell, and the comfortable middle is the most exposed place to stand.** Pick an end and move

toward it on purpose, while you still set your own price.

- **Abundance destroys trust, which makes trust the largest unbuilt infrastructure of the shift** – whoever owns what verifies the real captures the value of everything routed through it.
- **The new divide isn't access; it's depth — users (~90%), operators (~9%), architects (~1%) – and the 1% who build on AI as a platform take the overwhelming majority of the value.** The climb from user to architect is open to anyone willing to make it.

Finisher Move: Run a line through your own revenue and label each item "scarce-intelligence rent" or "real moat." The first list is what the clock is coming for on a schedule you don't control. Pick the single highest-margin item on it and start building its replacement this quarter, before the market reprices it for you.

Boardroom Question: *"Which of our current revenue lines is priced on the assumption that intelligence stays expensive – and what is our plan to replace that revenue before the market does it for us?"*

Sources & Currency Note

This book states facts as of mid-2026; specific figures move, the pattern does not. Key data points in this chapter: DeepSeek's disclosed training cost (~$5.6M – the compute for its V3 base model, not R1's all-in cost – Jan 2025) and the resulting ~$589B single-day NVIDIA loss (Jan 27, 2025 – largest one-day loss by a single company in market history); ChatGPT weekly active users (more than 900M by May 2026 and still approaching ~1B, OpenAI); McKinsey's estimate that AI could add ~$13T to global GDP by 2030 (2018 MGI – the book argues this undercounts a platform shift); GitHub Copilot generating on the order of 46% of code among active users (2025); YouTube uploads of ~500 hours of video per minute (2025); and the Arup deepfake fraud (~$25.6M, reported early 2024). Voice-cloning audio requirements ("a few seconds") and developer-population figures ("twenty-some million professional developers") are framed as ranges because published *estimates vary by definition and provider. Stage windows (2020–2023, 2023–2025, 2026, 2027+) are directional, not precise. Where a single figure could age badly, it is framed as a range or dated explicitly.*

CHAPTER 6

THE BATTLE FOR AI INFRASTRUCTURE

> **IN THIS CHAPTER** how the first company ever to cross five trillion dollars was built by pouring the toll booths — and how the giant that owned the road before it watched, understood, and still lost. **Why it matters:** in every platform shift, the infrastructure layer captures the overwhelming majority of the value, and the layer is being poured *right now*, in public. **What to do with it:** stop asking which AI vendor to buy from, and find the layer in your own industry that everyone else will be forced to route through — then decide, this quarter, whether you own it or rent it.

THE COMPUTE TOLL LAYER

Every workload in the economy routes through the same accelerator substrate — and pays the toll.

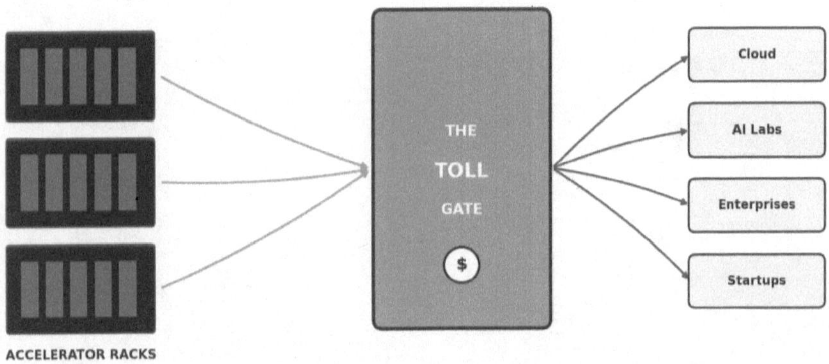

Own the gate, and you tax the entire revolution — whoever wins the apps.

There is a moment in every platform shift when the future stops being an opinion and becomes a bill. For the seventh turn, that moment is now, and it is being itemized in front of you: in market caps, in capital budgets, in restarted power plants. This chapter is about the layer where that bill gets written. Not the apps. Not the chatbots. The substrate underneath them: the silicon, the software that commands it, the data centers, and the electricity that feeds the whole thing. Stage 2 of the **Durable Sequence**, the Infrastructure stage, is where the toll booths get poured. We are standing in it. And the company that has so far poured the most is now worth more than any company in history.

Hold the introduction's five stages in your head as a single fact. Invention earns a footnote; infrastructure earns the century. This chapter is the proof, told through the cleanest example the modern economy has produced: a hardware company that understood it was actually a platform, and the king it dethroned, who understood the exact same thing thirty years earlier and could not act on it when the dynamic turned. Read it not as corporate history. Read it as your own org chart, with the names changed.

The Bet That Looked Like a Breakdown

It is 2015. Jensen Huang is on a stage in San Jose, addressing thousands of developers, researchers, and customers at his company's GTC conference, and behind him a slide says the company is betting everything on artificial intelligence.

Half the room thinks he has lost his mind, and that half is not wrong to think it. These are people who built careers, and mortgages, and families, on optimizing rendering pipelines for video games. NVIDIA was a graphics-card company, worth on the order of fifteen billion dollars: respectable, profitable, and a rounding error next to the semiconductor giants. The market its CEO was now telling them to chase generated, that morning, essentially zero revenue. There was no spreadsheet that justified the pivot, no analyst note that said *bet a profitable company on a technology academic researchers can barely get to work*. Some NVIDIA engineers updated their résumés within the week. The ones who stayed did it on something closer to faith than evidence, and a few of them are billionaires now. But in that room, in that moment, it did not feel like strategy. It felt like a founder's pitch, where the man at the front is either a visionary or a lunatic and you will not know which for a decade.

Now hold that fifteen-billion-dollar company next to the king it was supposedly challenging. Intel, as of 2015, was the undisputed sovereign of computing: worth roughly ten times more, with an order of magnitude more revenue, more engineers, more manufacturing, more of everything that a rational analyst counts. The Intel architecture was the universal standard. *Intel Inside* was not a slogan; it was a description of physical reality, printed on the case of nearly every computer on Earth. The idea that the graphics-card company could come for the throne was, by the arithmetic of 2015, a go-kart manufacturer announcing it would take on Toyota.

Most chief executives in Huang's position would have hedged. They would have stood up a division, fenced off ten percent of R&D, written "AI strategy" into the annual report, and gone back to defending the franchise. **That hedge is the single most expensive instinct in business, and it has a name later in this chapter.** Huang did not hedge. He turned the whole company toward the target, because he had internalized the one rule about platform shifts that Intel's leadership recited and never obeyed: *you commit completely, or you lose completely*. There is no defensible middle.

Eleven years later, the company that looked insane had become the first ever past $5 trillion. The king is a fraction of its former self, shedding share, shedding people, fighting to stay whole. That is not a parable about chips. It is the **Durable Sequence** running at full speed and the **Success Trap** closing in real time, and it is the clearest illustration ever recorded of why building for the future beats protecting the past — every time, without one exception.

The Fork: A Hundred Times, Not Ten Percent

The two roads forked at an academic contest that semiconductor executives had every reason to ignore.

In 2012, a deep neural network called **AlexNet** won the ImageNet competition, the task of correctly labeling objects in images, and it did so trained on NVIDIA graphics chips, not Intel processors. It cut the field's best error rate from about twenty-six percent to about fifteen, a margin nobody had come close to. And the reason it could be trained at all was the hardware. Not ten percent faster. Not twice. On the order of a hundred times faster.

Understand why, because the why is the whole strategy. A central processing unit, Intel's product, is built for sequential work: one complex calculation after another, executed brilliantly fast. Picture the world's finest mathematician, unbeatable at solving a single equation at a time. A graphics processing unit was built for the opposite: thousands of simple operations at once, because rendering a game means coloring millions of pixels simultaneously. Picture a stadium of average students. No one of them is brilliant, but together they finish a million simple problems in the time the lone genius finishes one. Training a neural network turns out to demand exactly the stadium: not one perfect calculation but billions of adequate ones, all at once. The GPU was accidentally perfect for AI. The CPU was structurally unsuited to it. The fork was not a product gap. It was a mismatch between an architecture and the shape of the future.

Two companies looked at the same hundred-fold result. Watch what each did with it, because the divergence is the lesson, and the lesson is not about silicon.

Intel's response was dismissal born of dominance. They were the king; AI looked like a hamlet at the edge of the kingdom. Their own analysts, by reliable accounts, sized the entire AI opportunity at roughly a billion dollars within a few years, and the real number arrived hundreds of times larger. That is not a forecasting miss. It is willful blindness in the service of a comfortable story: *our market is not under threat.* It was Kodak tallying film inventory while sitting on the digital camera it invented. It was Western Union counting telegraph poles while Bell wired the cities.

NVIDIA's response was total commitment. Huang did not bolt AI onto the product line; he aimed the entire company at it. He poured years of investment into CUDA, the software layer first released back in 2007 that turned NVIDIA's chips into general-purpose parallel machines. He built dedicated AI training systems and a full software stack on top of the silicon. And he hired software engineers until, by NVIDIA's own account, they made up the clear majority of the workforce, roughly eighty percent.

Sit with that last ratio, because it is the tell. A *hardware* company staffed mostly with *software* engineers. The other hardware company stayed mostly hardware engineers. That single number told you, in 2015, which firm understood what its own product had become and which did not. Intel believed it sold processors. It actually sold computing capability, and the capability was migrating to a form it did not control. NVIDIA believed, eventually, that it sold not chips but a platform for accelerated computing, of which AI would become the largest application in history. "We sell chips" versus "we sell the platform everything routes through" sounds like a semantic nicety. It is the difference between five trillion dollars and a twentieth of it.

The architect's takeaway does not depend on you being in the chip business. **The company that wins a platform shift is not the one with the best product. It is the one that correctly identifies what its product actually is.** Before you read another paragraph, answer it for your own company. Are you selling the thing on the invoice, or the layer the rest of your industry is quietly being forced to build on top of? Almost everyone gets this wrong, and the ones who get it wrong are, in the chapters of history, named after their products.

The Ecosystem That Became a Moat

CUDA is the most consequential piece of software most people have never heard of, and it is the part of this story you should steal.

Released in 2007, most of a decade before the AI boom, CUDA made NVIDIA's GPUs programmable for general-purpose math, not just graphics. Prescient or lucky; probably both. The result was that by the time the world's AI researchers needed massive parallel compute, the toolchain to wield it was already mature, refined, and everywhere. The hardware was the visible asset. The ecosystem was the moat.

Crypto Was the Dress Rehearsal

The compute story did not begin with cryptocurrency, but crypto taught a lesson worth keeping. Once CUDA made GPUs programmable, crypto did something the software alone could not: it created a global market that paid, immediately and in cash, for raw parallel throughput. By late 2010 the first publicly available GPU miners appeared, and within months "mining" stopped being software and became hardware economics: throughput per dollar, throughput per watt.

Then the ecosystem did what ecosystems always do once a workload stabilizes. It specialized and hardened. FPGAs arrived, then purpose-built ASIC miners, and by 2013 the CPU-to-GPU era of Bitcoin was effectively over. The workload froze, so the silicon froze around it.

What that boom left behind mattered more than the coins: a working prototype of compute-infrastructure culture. Operators stacked GPUs into rigs and racks and slammed straight into the exact constraints AI clusters hit today. Power delivery, cooling, availability, and the unglamorous truth that the bottleneck was never the idea, it was the hardware. One documented homebrew rig from that era ran dozens of GPUs cooled by box fans: an accidental preview of what happens when parallel compute meets warehouse physics. Crypto also proved that GPU demand could distort entire revenue streams, with gaming shortages, price spikes, and regulators asking how much "gaming demand" was really mining. The dress rehearsal had everything except permanence.

Then AI arrived and made the demand permanent. **Crypto was cyclical. AI is structural.** That single distinction is why one of these reshaped a product line and the other reshaped the economy.

And here is the same pattern at a higher altitude. Today's AI accelerators, priced in the tens of thousands of dollars apiece, increasingly behave like the ASIC miners that ended crypto's GPU era. They are bought like production equipment, racked in farms, governed by return-on-investment math and power ceilings, and packed with blocks tuned for one dominant workload: transformers. Technically they remain programmable accelerators, not fixed-function miners. But the direction of travel is identical, from CPU to GPU to specialization, and the specialization is already here in the form of custom AI silicon. The infrastructure war was never only *who has the GPUs*. It is *who owns the platform layer for compute itself*: the tools, the interconnect, the software, the ecosystem. Crypto was the rehearsal. This is the main event.

Why The Ecosystem Beat The Chip

CUDA outweighs any single piece of silicon for one reason. A generation of AI researchers learned to build on it. The breakthroughs were built on it. Every major framework was optimized for it. Every library, every tutorial, every forum answer assumed it. NVIDIA grasped what Intel never did: **developers, not hardware, decide the winner.** So NVIDIA invested in developers like the franchise depended on it: free university courses, research grants to promising labs, conference sponsorships that put the brand in front of every researcher in the field, open-source tooling that made the easy path the NVIDIA path. By the time those researchers walked into industry and their employers needed to buy AI hardware, they did not evaluate options. They *insisted* on NVIDIA, not always because the chips were fastest, but because switching meant rewriting mountains of code, retraining whole teams, and abandoning a decade of accumulated tooling.

This is platform economics in its purest form, so name it as law: **the best technology does not win — the best ecosystem wins.** NVIDIA built an ecosystem that made its hardware the only rational choice regardless of raw benchmarks. The rhyme with computing's last era is exact. Windows once made Intel's processors the default for a generation, the "Wintel" pact that ran personal computing. CUDA made NVIDIA's GPUs the default for AI. The cruel irony is that Intel had been the beneficiary of precisely this dynamic for thirty years and still failed to recognize it when the same mechanism turned and pointed the other way.

And NVIDIA did not stop at one layer. It stacked an entire software ladder above the chip: libraries for deep learning, optimizers for serving models, toolkits for data science and simulation, and each rung made the ecosystem deeper, switching costs higher, the platform more essential. That is the Durable Sequence playing out inside software: invention, then infrastructure, then platformization, then lock-in so complete that a competitor could ship superior silicon and still lose. Intel eventually launched an open alternative to CUDA. By credible accounts it was technically sound. Almost no one adopted it, because **you cannot unseat a platform with a product — you can only unseat it with a better platform**, and by the time Intel tried, CUDA had a decade-plus head start and the developers were already locked in.

If you build infrastructure, that is the move, stated as a build order. Ship the substrate early, give the tools away, make your path the path of least resistance, and let the switching cost compound while your competitors are still admiring their benchmarks.

The Product Ladder Intel Could Not Climb

While the king's strategy floundered, the challenger's product cadence revealed a company that knew exactly where computing was heading, because it was listening to everyone heading there.

Generation after generation, NVIDIA's accelerators delivered step-change gains aimed precisely at AI's evolving appetite: early chips for the first adopters; faster ones that pulled in researchers at scale; an enterprise standard that trained the first models consumers actually noticed; then chips built explicitly for large language models, each adding more memory bandwidth as models grew, faster interconnects as training spread across thousands of chips, and specialized matrix-multiplication hardware as the workload demanded it. NVIDIA was not chasing the market. It was building the market and then selling the market its next requirement.

The responsiveness was not magic, and it is reproducible. NVIDIA was in daily conversation with effectively every serious AI lab on Earth, because those labs all ran on CUDA. The developer ecosystem was not only a moat; it was an intelligence network with near-perfect visibility into where demand was going next. NVIDIA did not have to guess what researchers would need. It was told. Intel, without that relationship, was designing chips for a market it could not see clearly, and it is brutally hard to build the right product while guessing at the spec, and almost easy when your customers hand you the spec daily.

That is what total commitment buys. Not a side project, not a division, not a small team with a modest budget, but the entire company aimed at one target, with a free intelligence feed pointed at the future. **If you take one operational principle from NVIDIA, take this: the ecosystem that locks customers in is the same ecosystem that tells you what to build next. Lock-in and foresight are the same asset, bought with the same investment.**

The Cascade: How a Genius Company Loses

Intel did not fall on one bad decision. It fell down a staircase of individually rational decisions that were collectively fatal, which is the only way a company this good ever falls.

It bought an AI-chip startup in 2016 for hundreds of millions, promising a CUDA-class competitor; the products were cancelled before they shipped. It acquired the Israeli AI-chip company Habana Labs in December 2019 for about **two billion dollars**, billed internally as the "NVIDIA killer," and it captured a sliver of the market. It tried to re-enter graphics, pivoted to gaming GPUs few wanted, launched AI accelerators into a market where the ecosystem to use them did not exist. Billions spent; little to show. The throughline is the diagnosis. Intel kept its AI work separate from the core, staffed it lighter, funded it under the microscope of CPU-era margins, and graded every initiative against the profitability of the business it was supposed to disrupt. Every AI effort was an experiment running inside a company that was still, in its bones, a CPU company.

Then there was the manufacturing reversal, the part that should frighten any leader whose moat is operational excellence. Intel's edge for decades was that it made its own chips in its own fabs with processes rivals could not match. It fell behind anyway, while the contract manufacturer it once towered over reached the leading edge and Intel's own roadmap slipped and rebranded to disguise the gap. It lost the manufacturing edge and the design edge at once, and was left with neither. The firm that taught the world to fabricate chips ended up behind the firm that fabricates chips for its rivals.

The challenger had no such staircase, because it had no AI division to wall off. **The entire company was the AI division.** Every roadmap, every hire, every research dollar pointed at the same target. That is the difference between experimenting with the future and committing to it, and it is the difference, over a decade, between leading an industry and petitioning to survive in it.

The Earthquake — And the Gold Rush It Triggered

November 30, 2022, changed the clock for everyone.

ChatGPT launched and set an adoption record no technology had touched, the introduction's anchor: more than nine hundred million weekly users by May 2026, still approaching a billion. Overnight, the appetite for AI compute went vertical. Suddenly every company on Earth needed accelerators, not next year, not when the budget cleared, *now*, and demand did not climb by percentages, it climbed by orders of magnitude. For a window, there was one supplier that mattered.

The desperation was real and occasionally absurd. Chief executives of household-name companies called Huang personally to beg for allocation. Startups raised hundreds of millions where the pitch was not the product but the access to chips. Cloud providers hoarded silicon like bullion. A gray market emerged with top accelerators reportedly changing hands at multiples of list price. Companies built entire strategies not around customer demand but around how many chips they could secure. Waiting lists stretched past a year, prices rose, and the supplier could name its terms. It was not merely selling hardware. For a moment it was a kingmaker, deciding which companies could compete in AI and which could not.

For Intel, difficult became catastrophic. Budgets that had flowed to CPUs for decades swung to GPUs in a quarter. Data-center buyers who had bought Intel reflexively for a generation suddenly cared only about accelerators Intel did not have. The stock fell. Layoffs ran to the thousands. Talk began of breaking up the company, the ultimate confession that the whole had become worth less than the parts. The firm that had defined computing for half a century was, abruptly, peripheral to computing's future.

And here is the irony that should keep every dominant leader awake. Intel had the resources to build everything the challenger built: more engineers, more capital, more fabs, more customer relationships, and well over a decade of warning, from CUDA's 2007 launch to the late-2010s scramble. It did not choose blindness. It was *trapped*. Which brings us to the mechanism, because this is not a story about one company's bad luck. It is a story about a force that hunts the strong specifically because they are strong.

The Reversal, Honestly Stated

The market-cap arc reads like fiction, so let me state it the way it can be defended rather than the way it tempts you to exaggerate.

In 2015, the king was worth roughly ten times the challenger. Within a few years the gap had compressed to under two-to-one. Around the turn of the decade, the lines crossed: the go-kart passed Toyota. Then the curve went non-linear. The challenger pulled to several times the king's value, then an order of magnitude beyond it. In October 2025, **NVIDIA became the first company to cross five trillion dollars.** By mid-2026, Intel, recovered somewhat under new leadership from its lows, sits at a small fraction of that, well under a twentieth of NVIDIA's value.

In under a decade the positions did not merely reverse. They reversed and then separated by something like twenty to one. The student did not pass the master. The student made the master a supplier — and even being worth more than everyone is not the finish line; but that's a story for the next book.

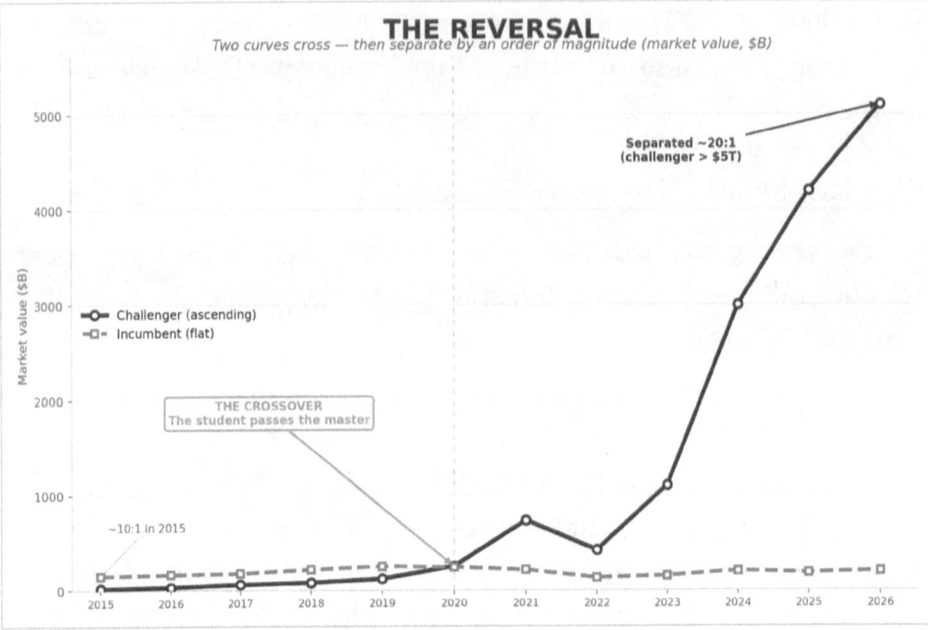

Intel's Partial Recovery — And Why It Is Not Enough

There is a coda, and it earns telling because it complicates the moral in a useful direction.

Under new leadership, Intel has shown a pulse. The stock has climbed substantially off its bottom. It has, in the cleanest possible symbol of who won, partnered with NVIDIA itself, taking a multi-billion-dollar equity investment and co-designing x86 products with the very company that beat it, turning the rival's dominance into its own lifeline rather than fighting it head-on. It has launched new AI inference silicon and debuted a next-generation manufacturing process. The foundry pivot, making chips for others, is a wholesale rethinking of what Intel even is.

These are real achievements, and they prove something important about the **Success Trap**: it can be partially escaped, with new leadership, radical strategic change, and enormous pain. But Intel at its recovered value is still a small fraction of NVIDIA, and the gap is widening, not closing. The recovery is survival, not restoration. The company that once defined computing now hopes to endure by building x86 processors for the company that defines computing's future. That is not a comeback. It is an obituary with a hopeful postscript.

The lesson is not that Intel is doomed. It is that **the time to move during a platform shift is before the shift, not after it.** NVIDIA started building toward this in 2007. Intel started scrambling more than a decade later. Those years were the entire difference between a five-trillion-dollar empire and a survival story.

And there is a warning here aimed squarely at NVIDIA, the same one that has come for every platform winner in history. **Success carries the seed of its own undoing.** If you want the template for what happens when one company controls essential infrastructure, look at AT&T. It did not lose to a better telephone. It lost because governments decided the platform had become too central for any single corporation to own, and broke it up. The brand survived only because a smaller offshoot later bought the carcass and took the name; the continuity is cosmetic. The real message underneath it is permanent: **platform winners do not only face competitors. They face regulators, reorganizations, and forced rewrites of the rules.**

Those forces are already gathering around the new infrastructure layer. Antitrust scrutiny is intensifying as governments register that one company sits astride the foundation of the most transformative technology since electricity. Customer concentration cuts both ways: if the largest cloud builders succeed with their own custom silicon, the dominant supplier's revenue could erode from inside its biggest accounts. And the DeepSeek shock from Chapter 4 proved efficiency can ambush the whole demand curve overnight. A single cheaper-than-expected model erased roughly **$589 billion** of NVIDIA's value in one trading day, the largest single-day loss by a single company in stock-market history. That was not a crash. It was the market repricing the cost of intelligence toward zero in a single session, and a live demonstration that no infrastructure monopoly is as permanent as it looks from the top.

So the question is not *whether* someone challenges NVIDIA. It is *when*, and whether NVIDIA, when the challenge lands, answers like the NVIDIA of 2015, with total commitment to the next shift, or like the Intel of 2015, defending the position it holds. That single choice will decide whether it becomes the AT&T of this era, dominant for a generation, then unbuilt by forces above the market, or escapes the pattern entirely. To his credit, Huang appears to understand the trap he is standing on. By his own account he told his people that the greatest danger is believing you have already won.

What Killed Intel — The Five Fatal Mechanisms

Chapter 1 named the force that hunts incumbents the **Success Trap**, and Chapter 3 took its five mechanisms apart. Intel triggered all five. This is not a recap of the framework. It is the autopsy, mechanism by mechanism, so you can find each one operating in your own company before it operates on you.

The Identity Prison. Intel *was* a CPU company, branded, culturally, all the way down, on what went inside the machine. Its hiring, its engineering rituals, its institutional memory were all tuned for one architecture. When the platform shifted, it could not re-author its own identity fast enough. Asking Intel to become a GPU company was like asking Kodak to become a digital-imaging company: the old self was not on the org chart, it was in the walls.

The Organizational Antibodies. Every AI initiative inside Intel was attacked by the CPU body that surrounded it. Resources got pulled back to defend the core. The A-players worked on the next flagship processor; the AI projects got the B-team and a budget reviewed against CPU-era margins. The company's immune system did exactly what an immune system does, and methodically killed the cells that could have saved the organism.

The Customer Trap. Intel's biggest customers, the hyperscalers, were telling it they needed faster CPUs. Intel listened and delivered faster CPUs. Meanwhile those same customers were quietly buying NVIDIA accelerators for their AI work. Intel heard what its customers said and missed what they did. The customers themselves had not fully named the shift yet; they asked for better versions of the old world while already building the new one. **Listening to customers is not strategy. Watching what they actually buy is.**

The Financial Gravity Well. CPUs threw off enormous gross margins; early AI hardware did not. Every financial model said *protect the CPU business*, and every model was right, in the short term, and wrong about the only term that mattered. This is the cruelest mechanism of the trap: during a platform shift, the financially rational decision and the strategically fatal decision are the same decision. If your CFO's spreadsheet and your strategist's map disagree, the shift is already underway.

The Pattern of Denial. "GPUs are for graphics." "AI is a small market." "CPUs will adapt." "We have time." "Our customers still need us." Each line was delivered with textbook precision by brilliant, experienced executives who simply could not accept that the world was leaving without them.

Here is the part to carve into the wall, because it is the difference between this framework and the comforting version everyone else tells: **the Success Trap does not catch stupid people. It catches smart people whose smartness is exquisitely optimized for a world that no longer exists.** Intel's leaders had decades of scar tissue and degrees from the best schools. They had navigated transition after transition, from mainframe to PC, PC to server, single-core to multi-core, and had every reason to believe they would navigate another. What they could not see was that this was not a transition *within* computing. It was a transition *of* computing: from sequential to parallel, from one architecture to another, from their foundation to someone else's. **You cannot navigate a transition that makes your foundation obsolete. You have to abandon the foundation and pour a new one — and that is the one act the Success Trap exists to prevent.**

The Broader War: Silicon, Cloud, and the Power to Run It

The Intel–NVIDIA duel is the sharpest illustration, but it is one front in a much larger war for the AI substrate, and the other fronts decide who controls the next era of computing.

Silicon is contested, not settled. NVIDIA dominates today, but dominance is a verb, not a possession. As of mid-2026, NVIDIA still commands the overwhelming majority of AI accelerators, on the order of eighty to ninety percent by most counts, but that share is expected to drift toward roughly three-quarters as alternatives scale, and the alternatives are not hobbyists. Google designs its own AI chips and trains its own models on them. Amazon builds custom silicon for training and inference to cut its dependence. Apple's in-house chips run AI with notable efficiency across a billion devices. AMD is pouring billions into competing accelerators and has a real toehold. And the model race itself now includes new entrants building enormous compute clusters as a wedge. Every major technology company is spending billions to build an alternative to the dominant supplier, and the Success Trap waits behind all of them, the same force that took Intel, ready for whoever stops running.

Cloud is the utility layer. A handful of hyperscalers, led by AWS, then Microsoft and Google, together more than sixty percent of global cloud infrastructure as of early 2026 (roughly thirty, twenty-four, and thirteen percent respectively, depending on whose methodology you trust), are building the data centers, laying the cables, and signing the power deals that will run AI for the next decade. They are the electric utilities of this era, and the introduction's iron law applies without amendment: whoever controls the infrastructure captures the lion's share of the value.

Energy is the foundation beneath the foundation, and the part almost no one priced in. Training and running AI consumes electricity at a scale that is now a strategic variable rather than an operating-cost footnote. Frame the numbers honestly, because the honest version is alarming enough. By the International Energy Agency's accounting, data centers used on the order of one and a half percent of the world's electricity in 2024, and that demand is set to roughly double by 2030 toward about three percent of global supply, with AI as the primary driver and AI-specific data-center demand growing several times faster than the rest. In the advanced economies where the build-out concentrates, data centers are projected to account for over a fifth of all electricity-demand growth through 2030. By common (and contested) estimates, a single large model's training run can consume as much power as a small town does in a year, and answering an AI query can take meaningfully more energy than a traditional web search. Hold the exact multipliers loosely; hold the direction tightly. The curve is exponential and the grid is not.

That curve has forced the technology giants into the energy business as an existential necessity, not a press release. They are signing deals to restart retired nuclear plants. The introduction's anchor here is **Three Mile Island**, reborn as the Crane Clean Energy Center under a twenty-year Microsoft–Constellation agreement announced in September 2024, with the restart of its 835-megawatt unit *now targeted for 2027*, and they are funding small modular reactors and geothermal at scale. The logic is a chain with no slack: without dedicated power, no data centers; without data centers, no trained or served models; without models, the platform war is lost. **You can always train a better model next year. You**

cannot train any model if you cannot power the building. The companies that lock up cheap, reliable energy will hold a structural advantage that no amount of algorithmic brilliance can overcome. Energy is the foundation beneath the foundation, and it is being claimed right now, quietly, in twenty-year contracts.

The scale of the build-out has no precedent in computing history. These are not the server closets of the early internet. They are hyperscale facilities each drawing more than a hundred megawatts, the electricity of tens of thousands of homes, sited for exactly three things: cheap reliable power, cooling, and network proximity. Individual operators are planning deployments measured in the hundreds of thousands of accelerators, and total investment in AI computing infrastructure is widely projected to run into the hundreds of billions of dollars within the next few years. This is the Infrastructure stage of the Durable Sequence pouring in real time, and the stakes are measured in trillions.

The whole war is happening in the background, invisible to the public, who meet AI through a chat box and never picture the silicon, the data centers, and the power plants behind the curtain. But the entities that control those resources will control AI itself, exactly as the owners of the power plants controlled the electrical age and the owners of the servers controlled the internet age.

> PLATFORM RULE — CARVE IT IN Whoever controls the infrastructure controls the platform. Whoever controls the platform captures the value. The electric utility decided which factories could run. The dominant mobile store decided which app businesses could exist. The AI substrate will decide which companies thrive. Your only real question is whether you will *own* a piece of that substrate — or *rent* it for the rest of your working life.

And rent is the right word, deliberately chosen, because it names what most companies actually are in this era. Every business building applications on top of someone else's AI infrastructure is a **sharecropper on a platform it does not own**: working land it cannot buy, with tools it cannot control, at prices it cannot negotiate. The day the platform owner decides to compete with it, raise the price, or rewrite the rules, it is finished. That is not theory. It is the lived history of everyone who built their company on a social platform's reach, an API that got revoked, or a search algorithm that changed overnight. The platform giveth. The platform taketh away. **If your entire business sits one pricing-page update away from extinction, you do not have a business — you have a tenancy.**

So here is the strategic imperative from this whole front, stated as a directive rather than an observation: **find where the infrastructure layer of your industry is being poured, and either own a piece of it or embed yourself so deeply into it that you cannot be pried out. Everything else is renting.**

Your Intel-or-NVIDIA Moment

Every company, every professional, every industry will face its own version of this fork: a platform shift that threatens everything it has built. The questions that decide which side you land on are uncomfortable by design, and they have to be answered with evidence, not intention.

Are you protecting the old model the way Intel protected CPUs, or building the new one the way NVIDIA built for AI? Most organizations believe they are NVIDIA and behave like Intel: they talk innovation in the all-hands while routing the overwhelming share of resources to defending the existing business. The litmus test cuts through every slide deck. Look at where your best people spend Tuesday. If your strongest talent is optimizing the current business, you are Intel. If your strongest talent is building the future, you might be NVIDIA. The org chart does not lie even when the strategy memo does.

Are you dismissing the shift because today's market is small? Intel saw a hamlet and called it irrelevant; the hamlet became a metropolis hundreds of times larger. Every platform shift starts small. Small is not a verdict that it does not matter. Small is the signal that you are early, and early is when the positions are still open and the fortunes are still unmade.

Are you building an ecosystem that creates lock-in? NVIDIA's real moat was never the chip. It was CUDA and the developers raised on it. Hardware gets copied. Ecosystems do not. So ask it plainly: what would it actually cost your customers to leave you? If the honest answer is "not much," you do not have a moat. You have a head start, and head starts expire.

What is your software strategy? Hardware without software is a commodity. Software without hardware is a dependency. The two together are platform power, and of the two, software matters more, because software is what manufactures the lock-in that compounds. Intel bet that better chips would win and lost. Better ecosystems win. What software layer are you building above your core product, and is it making you more essential every quarter or merely more featured?

Most critically, are you willing to cannibalize yourself? NVIDIA sacrificed a comfortable, profitable position to chase one that did not yet exist. Intel tried to protect the old and build the new simultaneously and failed at both, because in a platform shift those two goals are not complementary. They are at war for the same scarce resource: your best people and your nerve. **Self-disruption is painful and survivable. Being disrupted by someone else is fatal.**

The whole story collapses to one proof. In a platform shift, the company holding *every* advantage — unlimited capital, market dominance, customer loyalty, brand, decades of engineering excellence — can lose to a company holding exactly *one*: the willingness to bet on the future instead of guarding the past. NVIDIA's valuation is the reward for getting that right. Intel's decline is the price of getting it wrong. And the speed of the reversal, from ten-to-one underdog to roughly twenty-to-one leader in under a decade, is the part that should reset your calendar, because it tells you how violently these decisions compound once the curve goes vertical.

But the deepest lesson is not about chips or clouds or power plants. It is about **commitment.** Huang did not stand up an innovation lab and leave the franchise untouched. He did not hire a Chief AI Officer and issue a release. He turned the entire company, majority software engineers, every roadmap aimed at one target, every hire filtered through a single question: *does this person accelerate our future?* That looked insane in 2015 and inevitable by 2026, and **the distance between insane and inevitable is exactly where the fortunes are made.** Intel did everything *except* commit. It experimented, acquired, announced, restructured, and the gap between "experimenting with the future" and "committing to it," which looks like nothing from the outside, turned out to be the gap between five trillion dollars and a fraction of it.

Now make it concrete in your own world, because the substrate battle in *your* industry is not over. It is just beginning, and someone is pouring its foundation this quarter. In healthcare, the infrastructure layer is the data platform that fuses patient records, diagnostics, and treatment protocols; whoever owns it will govern how AI transforms medicine. In law, it is the research-and-analysis platform; whoever builds the CUDA of legal AI captures the profession's value. In finance, it is the risk-and-transaction layer. In education, the adaptive-learning systems. **In every industry there is a substrate being poured right now, and whoever pours it becomes the NVIDIA of that field while everyone who hesitates becomes the Intel — still operating, still respected, and no longer in control of its own destiny.**

And then sit with the part that is not about companies at all, because it is about you. Right now, in your career and your role and the way you spend your hours, you are facing your own Intel-or-NVIDIA choice. You can be Intel. You have the experience, the expertise, the track record that earned your seat, and you can spend it all defending what already works, dismiss AI as overhyped or nod at it abstractly while changing nothing, attend the conference, applaud the keynote, and return to your desk to do exactly what you did yesterday. That play feels safe and rational. It felt safe and rational to Intel's executives too, right up until the morning it didn't. Or you can be NVIDIA. Look at what is coming without the corporate-newsletter sanitation, see the full disruptive shape of it, and bet, not with a side project or a weekend course you will not finish, but with real commitment, the kind that makes your colleagues quietly wonder whether you have lost it, the same way Huang's people wondered in 2015. The difference between the people who thrive through a platform shift and the people it crushes is not intelligence, resources, or credentials. **It is courage**: the courage to abandon what works before it stops working, to look foolish in the short term to be right in the long one, to bet on yourself while the odds still look uncertain. The odds always look uncertain at the beginning. That is not a flaw in the opportunity. That *is* the opportunity. If the future were obvious, everyone would already be standing in it, and there would be no advantage left in arriving early.

Is the NVIDIA of your industry you, or the competitor you have not heard of yet?

In the next chapter, we prove the pattern across three worlds that have no business behaving alike — healthcare, law, and markets — and you will see why it is already coming for yours.

Recap: The AI Shift So Far

What you now know: AI is not a productivity tool. It is the seventh great platform shift, and it is democratizing intelligence itself, the scarcest resource in human history. As of mid-2026 it sits between Stage 2 and Stage 3 of the Durable Sequence, with the infrastructure being poured at a scale that dwarfs anything in computing history, and the cost of intelligence falling toward zero in real time. The company that has poured the most of that infrastructure, NVIDIA, is this era's Edison, while Intel is its Western Union: the prior sovereign, technically brilliant, left behind because it could not abandon the foundation that made it great.

What you must decide: Will you build *on* the AI platform, or become a customer of whoever does? And have you named the infrastructure gap in your own industry, the Pearl Street Station equivalent that *somebody* will build whether you move or not?

What happens next: The chapters ahead turn the pattern into action: the specific moves, tools, and timelines to position yourself as a finisher before the window closes.

The five things that matter:

- AI now meets or exceeds average human performance across coding, analysis, legal research, medical diagnosis, and frontline support.
- The cost of intelligence is collapsing toward zero, igniting the same adoption explosion cheap electricity once did.
- McKinsey's ~$13 trillion estimate undercounts the shift, because platform shifts create new categories, not percentage improvements on the old ones.
- The Intel–NVIDIA case is the Success Trap in real time: smart executives, rational decisions, fatal outcome.
- Whoever controls the AI substrate captures the lion's share of the value. Everyone else rents.

Key Takeaways

- The infrastructure layer captures the overwhelming majority of value in every platform shift, and AI's is being poured *now*, in public, which is precisely why a position in it is still available to those who move.
- NVIDIA's real moat was never silicon. It was **CUDA and the developers raised on it**: proof that *the best ecosystem wins, not the best product*, and that lock-in and market foresight are the same asset bought with the same investment.

- Intel saw the shift clearly for over a decade and still lost, because the **Success Trap** runs on five mechanisms — Identity Prison, Organizational Antibodies, Customer Trap, Financial Gravity Well, Pattern of Denial — and during a shift the financially rational move and the strategically fatal move are the *same* move.
- The war has three fronts: silicon (contested, ~80–90% NVIDIA today, drifting toward ~75% as custom chips scale), cloud (the big three > 60% of the market), and **energy** (the foundation beneath the foundation: data-center power roughly doubling toward ~3% of global electricity by 2030 per the IEA, forcing twenty-year nuclear deals).
- Every company building *on* someone else's AI substrate is a **sharecropper**, one pricing change away from extinction. Own a piece of the layer, or embed so deep you cannot be pried out. Everything else is renting.

Finisher Move: Name the single infrastructure layer in your industry that AI is making indispensable — the data platform, the research engine, the transaction rail — then write down, in one sentence, whether you are *building* it or *renting* it. If you cannot answer in a sentence, you are renting and do not yet know it.

Boardroom Question: *"Look at where our best people actually spent last week — not what our strategy says. By that evidence, are we Intel or NVIDIA right now?"*

Sources & Currency Note

This book states facts as of mid-2026; specific figures move, the pattern does not. Key data points in this chapter: NVIDIA as the first company to cross a $5 trillion valuation (October 2025); NVIDIA's AI-accelerator share (~80–90% in 2024–2025, projected to drift toward ~75% by 2026 as custom silicon scales — industry estimates, methodology-dependent); DeepSeek's reported training cost (~$5.6M — the V3 base-model compute, not R1's all-in cost — disclosed Jan 2025) and the ~$589B single-day NVIDIA loss that followed (Jan 27, 2025 — largest one-day *loss by a single company in market history); AlexNet's 2012 ImageNet win on NVIDIA GPUs; ChatGPT's launch (Nov 30, 2022) and adoption (more than 900M weekly users by May 2026, still approaching ~1B, OpenAI); Intel's acquisition of Habana Labs (~$2B, Dec 2019); the Microsoft–Constellation deal to restart Three Mile Island / Crane Clean Energy Center Unit 1 (835 MW, announced Sept 2024, restart targeted for 2028, later accelerated toward 2027); cloud infrastructure share (AWS ~28–31%, Azure ~21–25%, Google Cloud ~12–14%, early 2026 – Synergy/Statista, methodology-dependent); and data-center electricity (~1.5% of global use in 2024, projected to roughly double toward ~3% by 2030, AI the primary driver – International Energy Agency, "Energy and AI," 2025). Per-generation chip specifications, year-by-year market caps, and total-investment projections are described directionally rather than asserted to false precision. Where a single figure could age badly, it is framed as a range, dated explicitly, or attributed.*

CHAPTER 7

THE PATTERN IS EVERYWHERE

IN THIS CHAPTER the same five-stage **Durable Sequence** running, right now, through medicine, law, and money – three of the most regulated, credentialed, and protected guilds on Earth. **Why it matters:** if the pattern can route around a licensed profession, it can route around yours, and the only variable left is your position on the clock. **What to do with it:** locate the three-step value migration already underway in your own field – and decide which layer you intend to own before someone else assigns it to you.

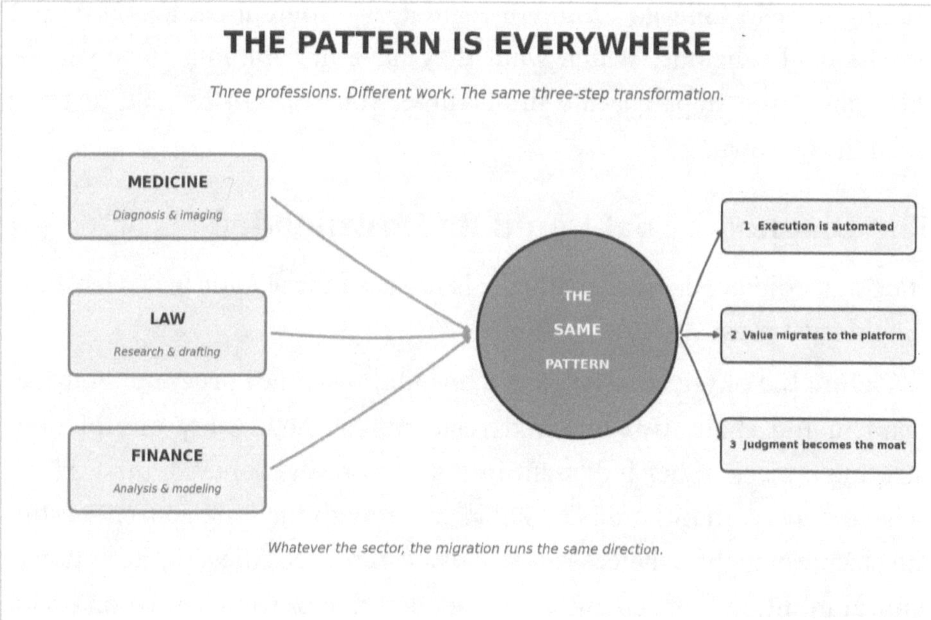

You have spent this book learning a single shape. The **Durable Sequence**: Invention, Infrastructure, Platformization, Proliferation, Consolidation. The inventor earns the footnote. The finisher who builds the infrastructure everyone is forced to route through earns the toll. Six times in a century and a half: manufacturing, computing, the internet, e-commerce, mobile, the cloud.

Here is the move this chapter makes, and why it matters more than any single case inside it. **The pattern does not respect the boundaries of an industry.** It is not running in "tech" while the serious professions watch from a safe distance. It is running inside the professions themselves, simultaneously, in the same quarter, on the same clock, and it is doing the most violence precisely where the credential was supposed to be the moat.

So we are going to walk into three rooms that look nothing alike. A hospital in the Midwest. A courthouse intake desk in the Southwest. A trading floor in Chicago. Different regulators, different cultures, different professional religions. Watch what they have in common. Once you see the same three steps execute in all three, you will not be able to unsee them in your own.

The Diagnosis That Could Be Downloaded

Start in medicine, because medicine is the case the skeptic trusts least and should trust most.

Picture the physician who runs a hospital residency program. Fourteen years in the chair, two hundred residents trained, every one of them carrying a piece of her judgment into a career. She believes, the way you believe gravity, that there is no shortcut through the long apprenticeship: the sleepless nights, the cases textbooks can't teach, the slow accretion of clinical intuition that separates a competent doctor from a great one. That belief is not vanity. For a century it was simply *true*. The repetitions could not be skipped, and so the person who had done them was scarce, and so she was valuable.

Then a diagnostic model gets put in front of the same ambiguous cases she uses to teach third-year residents how to think: overlapping conditions, incomplete histories, the deliberately hard ones. And it holds its own. Not on a lucky case. Across the set. Its near-misses are cogent near-misses, reasonable differentials that ranked the correct condition second instead of first, the same way a careful physician might. She asks to see its reasoning on the ones it got wrong, the way any rigorous teacher would. The reasoning is sound. That is the part that empties the room of air. Not that the machine guessed, but that it reasoned, legibly and defensibly, on the exact skill her career was built to certify.

This is no longer hypothetical, and you do not need an invented box score to feel it. **As of mid-2026, frontier clinical models score at or near the ceiling of the U.S. medical licensing exams that gate the profession.** One published model reported a score at or near the ceiling of board-style licensing exams in 2025. **And narrow diagnostic systems already match or exceed specialist sensitivity on specific reads** like pulmonary embolism on CT. The honest caveat, repeated by the same researchers and worth repeating here, is that exam performance is not bedside performance, and that the bottleneck to clinical use is no longer raw capability but implementation: workflow integration, monitoring, liability, trust. Hold that caveat. It is not a reason the pattern stops. It is a map of exactly where the next toll booth gets poured.

Because the question that should keep the residency director awake is not *"is the model good enough yet?"* It is structural, and it is the same question this whole book has been circling. **What does an apprenticeship sell when the baseline of competence it spent a decade installing can be downloaded in an afternoon?** The diagnostic value did not vanish. It is about to migrate, off the slow and expensive human pipeline and onto a platform that delivers it to anyone, anywhere, at the marginal cost of a query.

That is Step One. Hold it; we will name it in a moment, after two more rooms.

The Filing That the Guild Never Wrote

Now cross the country to a courthouse intake desk, and flip the disruption over, because in law it has a second face the profession would very much prefer you didn't see.

The standard story of AI and law is a partner-track story: billable hours compressed, associates anxious, BigLaw revenue under pressure. True, and covered. This is the other side of it. Consider the working parent with a custody matter and no money for it. The kind of person for whom a family lawyer at three to four hundred dollars an hour, across the twenty-to-forty hours such a matter takes, prices legal help at a full year's disposable income they do not have. Legal aid has a months-long waitlist. The clock on the filing does not.

So they do what tens of thousands of people are now doing: they describe the situation, in plain language, to an AI legal assistant. It asks the patient, specific questions a good intake attorney would ask, about the original agreement, the schedule, the record. It pulls the governing statutes. It surfaces the on-point local precedents. It drafts the response, the proposed plan, and the supporting affidavit, and it explains each section back in plain English on request. The filing that lands at the clerk's window is clean: properly formatted, correctly cited, procedurally sound. The kind of *pro se* filing court staff almost never see, because the people who file without lawyers are exactly the people who could never afford the craft that makes a filing look like that.

Under the human fact sits a structural one. **The well-documented access-to-justice gap means a large majority of Americans with a civil legal need navigate it without a lawyer:** eviction, custody, debt, immigration, the machinery of a life. For them, the legal system was never disrupted by AI, because it was never *available* to be disrupted. It is being *created* by AI. The guild that charged five hundred dollars an hour to fill in standardized forms is not losing its customers to a competing guild. It is being routed around entirely, and the market the technology serves best is the market the guild refused to serve at all.

Understand what that does to a defensive strategy. You cannot protect a position by being expensive when the disruptor's customers were the people your price already excluded. There is no moat against access. That is the second face of the pattern, and it is the one that ends professions rather than merely squeezing them.

The Curator of Machines

Third room. A trading floor, and the most important of the three, because this is the one that shows you the move, not just the wound.

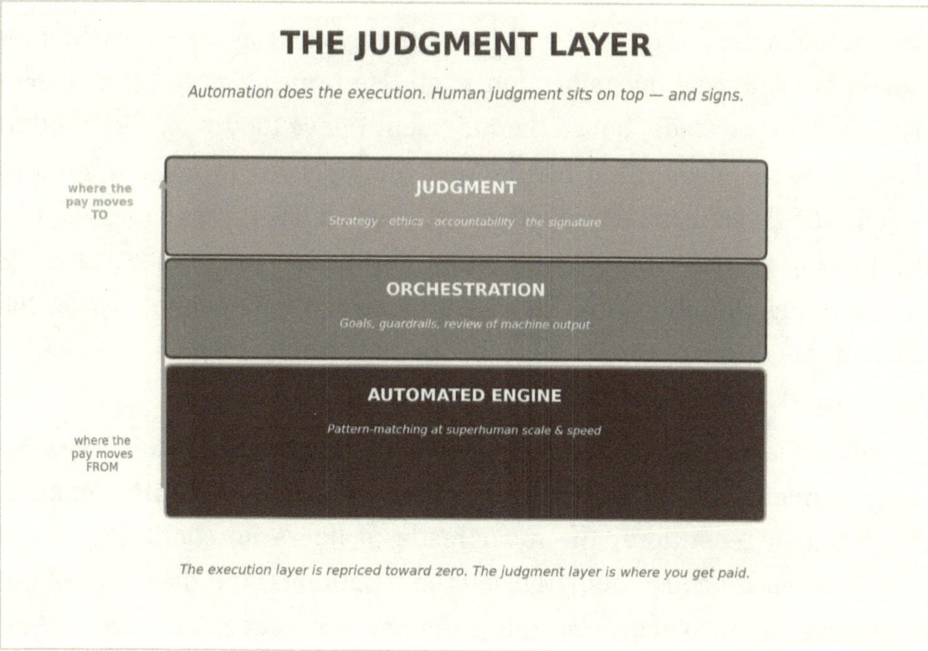

Picture the portfolio manager with two decades in the seat. Good, not legend, but reliably good, the kind who beat the benchmark more years than not, who could feel a sector rotating before the data confirmed it, who read an earnings call the way a card player reads a table. His firm stands up an AI trading system beside its human managers. A trial. A comparison. *Let's see how the machine does against the veterans.*

It beats him. It beats everyone in the firm, not on a hot quarter, across a year. Lower volatility, higher risk-adjusted return, faster rebalancing, no panic in a crash, no anchoring, no revenge trade after a bad week. He has watched this movie before, from a safe distance, in other people's industries. He expects to be fired.

He is not fired. He is *re-pointed.* The firm gives him a new charter, call it strategy curator, and his job is no longer to pick or to time. The machine does that better, and he is honest enough to say so out loud. His job is to evaluate the machines the way a curator evaluates work: which models carry a coherent thesis and which are overfit to the recent past, which price the black-swan tail and which are quietly correlated with each other and so offer no real diversification at all. He brings twenty years of scar tissue not to the trade, but to the judgment *above* the trade: the call that this strategy is elegant but fragile, that one ugly but robust. He decides which models get capital and which get shelved. He makes more than he did picking stocks, works fewer hours, and sleeps better, because he is finally doing the one thing the machine structurally cannot. He is the accountable human at the edge of an automated loop. The machine performs. It does not get to be responsible.

This is the formula this book has been building toward, made flesh. The old arithmetic of his career was **Value = Time × Skill**: seniority measured in repetitions, the moat made of hours no competitor could skip. The machine just industrialized the repetitions, and the floor fell out of raw execution. What replaced it is the new arithmetic: **Value = Intent × Leverage × Judgment.** He is paid now for the precision of his intent, the scale of the system he commands instead of the trades he places by hand, and the accountability of the judgment he signs his name under. He did not survive by fighting the machine and he did not survive by surrendering to it. He survived by moving up a layer, to the one place the value was migrating *toward.*

> FINISHER INSIGHT Man-versus-machine is the Dabbler's framing, and it loses both ways. Fire all the humans and you lose the judgment; ban the machine and you lose the speed. The Finisher's framing is *man-plus-machine versus everyone else.* You are not competing with the model. You are deciding whether you own the judgment layer that sits on top of it.

The Three-Step Pattern — Name It So It Can't Hide

Read the three rooms as one motion, because they are one motion. The pattern that ran through every one of them is the **Durable Sequence**, viewed from inside the professions instead of inside the data centers, and at the practitioner's altitude it expresses itself as three brutally consistent steps.

Step One — value migrates from practitioners to platforms. In all three rooms the intelligence did not shrink. The diagnosis took the same clinical reasoning; the filing took the same legal reasoning; the trade took the same market reasoning. What changed was the delivery mechanism. When delivery moves from a scarce, slow-trained human onto a platform that serves the same output to everyone at near-zero marginal cost, the economics don't improve incrementally. They reprice by orders of magnitude, the way the price of light collapsed once the grid was poured.

Step Two — the platform owner captures the leverage. The hospital is not building the diagnostic model; it is licensing one that will be licensed to ten thousand more hospitals, draining value out of the slow apprenticeship that used to mint diagnostic competence one resident at a time. The legal platform serves thousands at a subscription price, capturing a market that never flowed to anyone before, because it could not afford to exist. The trading system grows more valuable with every cycle it learns from, while each client firm needs fewer human hands. Different industries, identical mechanic: **the owner of the layer everyone routes through captures the value of everyone routing through it.** That sentence is the whole book in one line, and here it is executing three times in one chapter.

Step Three — finishers build on the platform and win. This is the only step you have a vote in. The hospital that re-architects care around the model, machine on triage and routine reads while humans concentrate on the complex and the relational and the irreducibly human work of healing, pulls ahead of the hospital that treated the model as a threat and now runs slower, costlier, and less accurate while patients walk. The firm that rebuilds around AI for research and drafting and re-points its lawyers toward strategy, negotiation, and advocacy serves more clients at lower cost and beats the holdout that has become more expensive than a solo practitioner with a subscription. The asset manager that creates the curator's seat, a judgment ring around a computational core, outperforms both the firm that fired all its humans and lost the judgment, and the firm that banned the machine and lost the speed.

Three steps. Three industries that share no regulator, no culture, no professional creed. One sequence, one outcome. This is not analogy. It is the same machine, running in three buildings at once.

If It's True in These Three Worlds, It's True in Yours

I could keep walking you through rooms: accounting, architecture, logistics, marketing, code. The examples are inexhaustible and, past a point, beside the point. **The examples are not the lesson. The pattern is the lesson**, and the pattern does not care what is printed on your office door.

Every knowledge-work industry on Earth is somewhere on the same curve, right now, simultaneously. The only variable is timing. Medicine is held at the edge of platformization by regulation and liability, Stage 2, with the toll booths visibly under construction. Finance is further along, already paying tolls and re-pointing its veterans. Law is being disrupted from the *outside* precisely because it refused to disrupt itself from within, the Dismisser's tax, collected in full. And the most advanced factories already run lights-out, because manufacturing got its electricity lesson a century ago and learned to redesign the plant around the new power source instead of bolting it onto the old one. That redesign-versus-bolt-on distinction is not a metaphor here. It is the literal difference between the 10x and the cost wall, and it is coming for the desk professions exactly as it came for the factory floor.

So locate yourself on the curve, honestly, the way the three types demand. Roughly nine in ten leaders read this chapter as *interesting* and change nothing: Dismissers, reasonable and skilled and on their way to becoming footnotes. Roughly nine in a hundred will spin up a pilot, appoint a Chief AI Officer, commission the deck, and protect the profitable old workflow so completely that AI never touches it: Dabblers, mistaking motion for migration. The remaining one in a hundred will do the only thing that has ever mattered in a platform shift. They identify the layer in their industry where value is migrating *toward,* and they move to own it before the window closes.

That window is short and you know its arithmetic by now: ignition in late 2022, roughly three and a half years elapsed as of mid-2026, the decisive phase compressed into the next twenty-four months. The pattern will reach your industry. That was never the question. **The only question is whether you will be a finisher or a footnote when it arrives**, and that is decided not by your talent, but by the date you decide to move.

The next chapter stops describing the pattern and starts arming you against it: the five executive moves that separate the companies that capture this shift from the ones it captures.

Key Takeaways

- Medicine, law, and finance share no regulator, culture, or professional creed, yet the **Durable Sequence** runs through all three on the same clock. If the pattern routes around a licensed guild, the credential was never the moat you thought it was.
- At the practitioner's altitude the sequence shows up as three steps: **(1)** value migrates from practitioners to platforms, **(2)** the platform owner captures the leverage, **(3)** finishers who build on the platform win. You only get a vote in Step Three.
- AI doesn't only steal a profession's customers. In law it *creates* a market the guild refused to serve. There is no defensive moat against access; you cannot out-price a disruptor whose customers your price already excluded.
- The curator's seat is the lesson: survival is not man-versus-machine but **man-plus-machine versus everyone else**, a thin ring of accountable human judgment around an automated core. That ring is **Value = Intent × Leverage × Judgment** made operational.
- Your industry is already on the curve; only the timing differs. Bolt AI onto the old workflow and you hit the cost wall; redesign the org around it and you get the 10x, the same lesson the factory floor learned from electricity a century ago.

Finisher Move: Take the single highest-margin task your people still perform by hand, and this week put its real output side by side with what a current AI platform produces at a fraction of the cost. Don't debate it in the abstract. Run the comparison and read the result the way a finisher reads a trench being dug.

Boardroom Question: *"Which of the services we charge the most for could a solo operator armed with today's AI deliver at comparable quality – and what does that do to our pricing, our headcount, and our moat?"*

Sources & Currency Note

This book states facts as of mid-2026; specific figures move, the pattern does not. Claims in this chapter are framed as dated or directional rather than precise where the underlying numbers are perishable: frontier clinical models scoring at or near the ceiling of U.S. medical licensing exams and matching specialist sensitivity on specific diagnostic reads (e.g., pulmonary-embolism CT detection) — with the documented caveat that exam and benchmark performance is not bedside performance and that implementation, not raw capability, is the current bottleneck; the well-documented U.S. access-to-justice gap, in which a large majority of Americans with a civil legal need proceed without a lawyer; and the compression of the platform-shift window (AI ignition late 2022, ~3.5 years elapsed by mid-2026, decisive phase through the next ~24 months). The three workplace scenes are composite archetypes drawn to illustrate the verified pattern, not reports of specific named individuals.

CHAPTER 8

THE EXECUTIVE MOVE-SET

IN THIS CHAPTER the five moves, in order, that turn everything you've read into something you can start Monday – and the three one-page instruments that score whether you actually did. **Why it matters:** strategy without a specific move-set is corporate theater, and the platform shift does not pay out for theater. **What to do with it:** pick your value-capture point before you touch a tool, run the 90-day scoreboard, and find out – honestly – whether you are the finisher or the customer.

THE EXECUTIVE MOVE-SET

Five moves, one rail, fixed order — each fails if you skip the one before it.

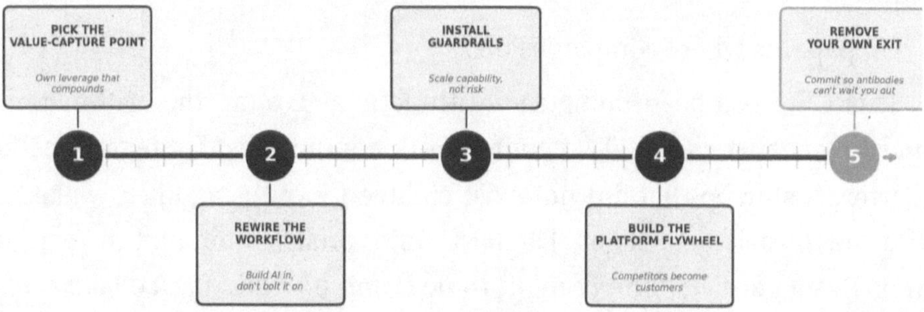

Skip Move One and you automate your way to parity. The sequence is the strategy.

Every chapter until now has been diagnosis. This one is the prescription.

You know the pattern. The **Durable Sequence**, five stages, the finisher who builds the infrastructure capturing the value the inventor only proved. You know the trap: the five mechanisms of the **Success Trap** that turn your strengths into the exact reasons you can't move. You know the clock is the shortest in history and the decisive phase is the next twenty-four months. None of that is worth a cent until it becomes a decision you make and a move you execute.

So let me hand you the move-set. Not ten moves. Not a maturity model with forty boxes. **Five moves, in a fixed order, each of which fails if you skip the one before it.** Then three instruments you can print and fill out in your next leadership meeting, because the gap between the executive who *agrees* with this book and the one who acts on it is not insight. It is a scoreboard and a start date.

Let me show you what executing the move-set actually looks like, run it on a single company at full speed, and then hand you the same controls.

Ninety Days, One Whiteboard

Picture the CEO of a two-hundred-person professional-services firm. Twelve straight profitable years. Steady revenue, decent margins, a balance sheet that gives no one any reason to panic. On paper, everything is fine.

On paper is where companies die.

This CEO has been doing the arithmetic at 2 a.m., the arithmetic no quarterly report shows. Two competitors announced AI integrations last quarter. A startup that did not exist eighteen months ago just walked off with the third-largest client. The best junior analyst quit, and the exit line lands like a verdict: *"I don't want to be doing by hand in five years what a machine already does in five minutes."* That sentence is the **Durable Sequence** speaking through a twenty-six-year-old. The execution layer is being repriced toward zero, and the talent can feel it before the income statement can.

So one Monday, fifteen minutes before anyone else arrives, three words go on the whiteboard: **AI Transformation — 90 Days.** The CFO calls it a waste. The CTO is afraid. The board wants results and can't agree on which kind. Underneath it all, a second line: *Transform or become irrelevant. There is no third option.*

Here is what does **not** happen next, and it is the whole lesson. No pilot. No steering committee. No three-hundred-thousand-dollar readiness deck destined to be read once and filed. Instead: a public, irrevocable commitment with a resignation on the table if it fails. That is not theater.

It is the one act the **Success Trap** cannot survive, a leader removing her own exit so the **Organizational Antibodies** have nothing to wait out.

Watch the ninety days run.

Day 30 — proof, not promises. Customer response times collapse from three days to half an hour. A pricing leak buried in the data for who-knows-how-long gets found and closed in an afternoon. The most fossilized skeptic in the building, fifteen years of resisting every system change, raises a hand and asks if she can try it on her team's reporting. That is the antibody converting. It is also the only kind of evidence that converts antibodies: a number with her name on it.

Day 38 — the trough. Two top account managers quit. A marquee client calls with "concerns." A board member texts a single sentence: *Are you sure about this?* The resignation letter gets drafted at 9 p.m. and deleted at midnight. Every real transformation has this day. The ones that fail have it and turn back. **Plan for the trough before you start, or the trough will make the decision for you.**

Day 61 — the reframe that changes the company. A new question goes on the whiteboard: *Do we want to be a company that uses AI, or the platform our industry runs on?* This is the chapter's hinge, the **Durable Sequence** applied to one's own firm. The methodology, the thing the company actually knew that no general model did, gets packaged as a product and licensed to smaller competitors who will never build their own. The competitors become customers. The customers generate data. The data sharpens the platform. The flywheel catches.

Day 90 — the new arithmetic, on the board. Costs down materially in the functions that were rebuilt. Throughput up several-fold where workflows were redesigned, not decorated. A new revenue line, platform licensing, that did not exist ninety days earlier and is compounding. The board votes to double down.

That CEO did not "adopt AI." She repositioned the firm from a service provider into the rails its competitors now rent. She made the finisher's move. And this is the part for you: none of it was about technology. It was **five decisions, made in the right order.** Those decisions are available to anyone with the nerve to remove their own exit. Here they are.

The Five Executive Moves

Every company that survives a platform shift makes the same five moves, and the order is not a suggestion. Skip Move One and you automate your way to parity. Skip Move Three and you scale risk faster than capability. The sequence *is* the strategy.

Move One: Pick the Value-Capture Point

Before you license a single tool, answer one question, and answer it on paper: **Where will you own leverage that compounds?**

Almost everyone opens with the wrong question, *"What can we automate?"*, and the wrong question routes you straight into the cost wall. Automation is a commodity. Whatever you automate, every competitor automates inside a year, the savings cancel out across the sector, and you have run a race where everyone crosses the line together. Worse: this is precisely the move the **Dabbler** makes. You know the MIT figure from Chapter 3: roughly **95% of corporate generative-AI pilots returned nothing to the bottom line**, not because the models were weak, but because organizations forced AI into existing workflows with minimal adaptation. "What can we automate?" is the question that produces that 95%.

The finisher asks the inverse: *Where in our value chain can we build a position that gets stronger with every transaction, every data point, every customer interaction?* The CEO's answer was a proprietary methodology that, fed by client data and sharpened by AI, became something no competitor could clone, because they lacked the data, the feedback loops, and the years of judgment baked in.

Yours will be specific to you. Your regulatory expertise. Your customer relationships. Your supply-chain intelligence. The one thing you know about your market that no generalist model trained on the open internet will ever know. **Find it before you start. Write it in a single sentence. Everything downstream — every tool, every hire, every workflow — is justified only by how directly it deepens that one position.** If a proposed AI project does not strengthen your value-capture point, it is someone else's strategy wearing your budget.

Move Two: Rewire the Workflow — Don't Bolt AI Onto It

The most expensive mistake in this entire shift is treating AI as a layer you add to a process built for humans. You take a workflow designed around human hands and bolt a model onto one step of it. That is a jet engine strapped to a horse cart. The cart was never built for the speed; it does not go faster, it comes apart.

This is not a metaphor. It is the literal mechanism behind the cost wall, and history already ran the experiment. When factories first electrified, owners bolted electric motors onto layouts designed around a single central steam engine, and for roughly **two decades productivity barely moved.** The gains, when they finally came, arrived only when plants were redesigned *around* electricity: machines placed by workflow instead of by proximity to a driveshaft. AI rhymes exactly. Bolted on, it is a line item. Built in, it is the 10x.

So don't automate the old process. Delete it and design the one you would build if you were starting the company today with AI available on day one. The CEO's client-onboarding flow touched fourteen people over nine days; the redesign collapsed it to under a day, not because a tool made the fourteen steps faster, but because eleven of them stopped existing. The instant a deal closed, the system provisioned accounts, configured the customization, scheduled training, matched a success manager on expertise, and generated the welcome package in parallel.

The discipline is one question, asked of every core workflow. Not "how do we do this faster?" but "if this company were founded this quarter, with these capabilities, how would this even work?" Redesign to that answer. **The companies winning the next thousand days are not running yesterday's processes faster.** They are building a thin ring of human judgment around an automated interior, and out-shipping firms ten times their size for the price of a few subscriptions. That is the new org chart. It pays the builder of the rails, not the doer of the steps.

Move Three: Build the Trust Layer — Before You Need It

This is the move nearly everyone skips, and it is the one that turns a fast start into a lawsuit.

When reality itself can be manufactured by deepfakes, hallucinations, synthetic content indistinguishable from the real thing, trust stops being a feeling and becomes infrastructure. Something built, verified, metered, and owned. Deploy AI into customer-facing work without that infrastructure and you are not moving fast; you are accumulating a liability that compounds silently until it arrives all at once.

The trust layer has three load-bearing components, and you build them as a system, not as a memo:

- **A verification gate.** Every AI output that touches a customer, a contract, or a consequential decision passes a human checkpoint until you have *earned* the confidence to widen it. Not forever, but until the data says the gate can open.
- **A provenance trail.** You can trace any AI-generated deliverable back to its inputs, the model that produced it, and the human who signed it. When a client or a regulator asks "how was this made and who stands behind it," you have an answer in seconds, not a war room.
- **A policy boundary.** Explicit, written rules for what AI may do autonomously, what requires a human in the loop, and what it is never permitted to touch. Drawn before the incident, not during it.

Do not mistake this for bureaucracy. **It is the infrastructure that lets you scale deployment without scaling risk in lockstep**, the thing that makes "move fast" survivable. And it maps directly onto the new arithmetic: the provenance trail is where **Judgment** lives. The machine performs; it does not get to be responsible. The signature of an accountable human at the edge of an automated loop is the one thing the model cannot supply, which is exactly why it is worth building a system around, and the thing your faster, sloppier competitor will be missing when the headline lands that gets a CEO fired.

Move Four: Compound the Data Advantage

Models are rented. Data is owned. Everything you can buy off a vendor's shelf, your competitor can buy off the same shelf by Friday. The only moat that widens while you sleep is proprietary data feeding a loop.

Most organizations discover, the moment they reach for it, that their advantage is fragmented into uselessness. The same customer living under different identifiers in eleven systems, the same product named five ways across five departments. The intelligence is there. It is just shattered. **Unify it first, and the models do not improve gradually; they step-change.**

What unification unlocks is not a faster report. It is a new sense. In the CEO's case, once the data was joined, it began to *predict*: client churn became visible roughly three months out, not from any single metric but from the combination of three quiet shifts. Declining login frequency, a cooling in the emotional tone of support tickets, and a drop-off in advanced-feature use. Each invisible alone. Together, an early-warning system no competitor could replicate, because no competitor had the joined history to learn it from.

That is the data flywheel, and it is the hardest moat in the shift to attack: more data sharpens the models, sharper models deliver better service, better service wins more customers, more customers generate more data. Every turn widens the gap. A rival who starts today does not start even. They start a full cycle behind, and the distance grows daily.

So instrument everything that moves, unify it under one identity spine, and protect it like the balance sheet it is. **Your data strategy is not an input to your business strategy. It is your business strategy.**

Move Five: Stand Up the Finisher Team — Permanent, Not a Project

AI transformations do not fail because the technology breaks. They fail because the organization doesn't, and organizations do not change without specific people whose entire job is to change them against the resistance of everyone optimizing for last year.

The CEO created three roles that had not existed. Someone who had already scaled AI at a serious company and could tell a real deployment from a demo. Someone whose libraries measurably lifted output quality across teams. And a data owner elevated out of the back office into a direct report, because data was now strategy and strategy reports to the top. The mandate attached to each role is the part most companies miss: **spend a fifth of your time teaching the people already here.** You are buying expertise to *manufacture* expertise, to multiply your existing staff, not to quietly replace them.

And the team is not a project with an end date. **Platform shifts do not end; they consolidate.** If your AI initiative has a finish line, you have already lost, because the firms that win are the ones with a standing capability whose only job is to keep pushing the boundary of what AI enables here, this quarter, forever. The project mindset is a Dabbler tell. The permanent-capability mindset is what a finisher builds. One ends; the other compounds.

The 90-Day Scoreboard

You do not need a calendar. You need a scoreboard, because "we're working on AI" is what a company says right up until it becomes a customer. Here is what *winning* reads like at each checkpoint. Print it. Track it. Be merciless about the gap.

By Day 30 — ground truth established:

- Every employee has produced *real* output with AI: shipped work, not a sandbox demo.
- Three measurable quick wins are live and generating results people can see.
- Your value-capture point is named and written in one sentence.
- Data unification has actually begun, not been scheduled.

By Day 60 — the machine takes shape:

- At least one core workflow is *redesigned*, not decorated.
- The trust layer is standing: verification gate, provenance trail, policy boundary.
- The data flywheel has produced its first real insight.
- At least one AI-powered offering is in pilot with real clients, not internal mock-ups.
- The **Organizational Antibodies** have been confronted directly, with no slow-rolling and no quiet veto.

By Day 90 — the new arithmetic shows up in the numbers:

- New revenue from AI-enhanced offerings is measurable, not theoretical.
- Cost is down by a meaningful margin in at least one major function, because you redesigned it, not because you trimmed it.
- The finisher team is hired and operating.
- Your moat, the position that strengthens with every transaction, is visible and deepening.

- You can state, in one sentence, what makes your position genuinely hard to replicate.

Hit these and you are a Finisher. Miss three or more and you are a Dabbler. And the **Dabbler does not lose to AI. The Dabbler loses to the Finisher who used these ninety days while you used them to deliberate.**

Three Instruments for Monday Morning

What follows are not concepts. They are one-page instruments built to be used in your next leadership meeting. Print them. Fill them out in the room. They will tell you where you stand and what to do before you leave it.

Instrument 1: The Executive Diagnostic

Where are you exposed, and where can you win?

Score your organization on five dimensions. The only person a soft score fools is you.

1. Workflow Redesign Readiness *(0–10)*

- **0–3:** Work is mostly manual; no AI in real use; processes undocumented.
- **4–6:** Some tools in use; key workflows documented; pilots underway.
- **7–10:** Core workflows redesigned around AI; measurable throughput gains; continuous optimization.

2. Data Infrastructure *(0–10)*

- **0–3:** Siloed data; no common identifier; basic questions take a week to answer.
- **4–6:** Partial unification; some real-time pipelines; can describe the past, not predict the future.
- **7–10:** Unified platform; real-time pipelines; models improving with every transaction.

3. Talent Pipeline *(0–10)*

- **0–3:** No internal AI expertise; fully vendor-dependent; staff anxious about AI.
- **4–6:** Some skilled hires; training in progress; uneven adoption.
- **7–10:** Standing finisher team; broad AI literacy; a culture that learns continuously.

4. Competitive Threat Timeline *(0–10)*

- **0–3:** Competitors already ahead; AI-native entrants courting your clients; no defensive position.
- **4–6:** Aware of the threat; some response underway; position uncertain.
- **7–10:** Leading the sector; competitors copying *your* moves; clear differentiation.

5. Platform Positioning *(0–10)*

- **0–3:** Pure service provider; value captured in labor, not systems; no platform play.
- **4–6:** Building toward platform; some recurring/scalable revenue; early network effects.
- **7–10:** Platform operational; competitors becoming customers; flywheel spinning.

Your total: ______ / 50

- **40–50 — Finisher.** Stay aggressive; the lead is real but the window still closes.
- **25–39 — Dabbler.** The window is open and closing. Move this quarter, not next.
- **10–24 — Dismisser with a spreadsheet.** The activity is real; the strategy is absent. Wake up.
- **0–9 — Kodak in 2005.** You invented nothing you couldn't have led with. The clock is loud.

> **DECISION POINT — IN THE ROOM, RIGHT NOW** Score the five. If your total is under 15, you are not the CEO who removed her own exit. You are the one across town still asking for "more data." The fix is not more analysis. It is a date on the whiteboard and a move you cannot quietly abandon.

Instrument 2: The Trust Ladder

Five rungs from "no AI policy" to "AI-native with a full audit trail."

Where are you standing, and where must you stand twelve months from now?

Rung 1 — No Policy. No formal usage policy. Employees using AI ad hoc, no guidelines, no oversight, no consistency. Where most of the market sat in 2024. If you are still here, you are exposed: to liability, to inconsistency, and to staff making commitments your organization cannot verify or defend.

Rung 2 — Guidelines Exist. A written policy, an approved-tool list, basic training done. But no mechanism to confirm anyone follows it. A start, not a system.

Rung 3 — Verification in Place. A human-review gate for every AI output touching clients, contracts, or critical decisions. Provenance tracking on deliverables. Clear escalation for edge cases. At this rung you can tell a client or a regulator exactly how a piece of work was produced and who signed it.

Rung 4 — Integrated Trust Architecture. AI and human workflows designed *together*, with trust checkpoints built in rather than bolted on. Automated monitoring flags anomalies before they reach a customer. Regular audits against human baselines. At this rung your AI deployment is more reliable than your pre-AI process ever was.

Rung 5 — AI-Native, Full Audit Trail. End-to-end provenance on every interaction. Continuous monitoring. Automated compliance. Client-facing transparency: customers can see how AI contributed, and they trust it because you built the infrastructure that earns trust. At this rung, your trust layer is no longer risk mitigation. It is a competitive advantage you can put in a sales deck.

Your current rung: ______ Required in 12 months: ______ Gap: ______ rungs

Each rung is roughly sixty to ninety days of focused work. A gap of three or more rungs means you are scaling on quicksand. Start this week, because the trust layer is the one piece you cannot retrofit after the incident.

Instrument 3: The Uncertainty Portfolio

Do not bet the company on one model, one vendor, or one architecture.

The landscape is moving faster than any technology in history, and the ledger proves it. Remember the DeepSeek day from Chapter 4: one cheaper-than-anyone-expected model knocked roughly **$589 billion** off NVIDIA in a single session, the largest one-day loss by a single company in stock-market history. That is what a repricing of intelligence looks like in real time. The model that's best this quarter may be mid-tier by year-end. The vendor that dominates now may be disrupted next. **Move aggressively, but build optionality into every move, because fragility is the one thing this market punishes faster than caution.**

Five deliberate hedges that preserve your freedom to move while you move:

1. Multi-Model Strategy. Don't marry one provider. Route different tasks to different models, keep sensitive data on models that never leave your infrastructure, and build an abstraction layer so you can swap engines without rewriting the application. Being locked to a single provider is catastrophic the day it changes pricing, capability, or its mind.

2. Vendor-Agnostic Data Layer. Your data outvalues any model. Hold it in systems you control, in formats any model can consume. If your advantage lives inside a vendor's proprietary store, you do not own your moat. *They* do. Build so you can plug in any AI without a migration.

3. Reversible Architecture. When you are uncertain, and you should be, about plenty right now, make reversible decisions. Choose architectures you can unwind cheaply. Pilot before you commit. Build modular, so a component can be replaced without rebuilding the whole. The companies that get hurt are not the ones that decide wrong; they are the ones that decide *irreversibly* wrong.

4. Talent Diversification. Don't staff entirely from one ecosystem. Your engineers should know more than one model; your architects should design for flexibility, not for a single vendor's strengths. A team that knows only one stack is one vendor pivot from starting over.

5. Compute Optionality. Assume the hardware stack shifts, from general-purpose GPUs toward specialized accelerators, the way every compute wave eventually has. Keep workloads portable, negotiate reversible capacity commitments, and preserve the ability to move between GPU-heavy, ASIC-heavy, and hybrid as costs and performance move. And do not confuse winning distribution with owning the platform: bundling a browser once won the desktop and still did not win the center of gravity of the web.

The Uncertainty Portfolio is not caution dressed up. **It is how you stay aggressive without becoming fragile**, the difference between betting big and betting the company.

The CEO Across Town

There is another CEO in the same city. Same articles. Same conferences. Same warnings. Not stupid, not uninformed, not in denial. He knows, intellectually, that AI is remaking his industry. And he is going to lose anyway, which makes him the more instructive case, because his failure is not ignorance. It is a series of reasonable-looking substitutes for the one thing that mattered.

He launched a pilot. Small team, limited scope, *"let's see how it goes before we commit."* It produced modest results: enough to justify itself, not enough to change anything. He declared it a success and moved on.

He formed a committee. Twelve people, six departments, monthly meetings, a forty-seven-page strategy document no one read, and a recommendation for "further study" on three promising initiatives still being studied today.

He hired a firm to assess his "AI readiness." Several hundred thousand dollars and four months later, a beautiful deck confirmed what he already knew, and went into a shared drive no one opens.

He is waiting. For more data. To see how competitors fare. For regulatory clarity. For proof that AI won't plateau before he bets on it. He is being prudent. He is being responsible. He is being careful.

He is going to become the other CEO's customer.

Not because she was smarter. She wasn't. Because she **decided.** She picked a value-capture point. She rewired a workflow. She built the trust layer. She compounded a data advantage. She stood up a finisher team. She made the five moves, in order, against the same fear he felt, and then she removed her own exit so the fear couldn't talk her out of it.

He made a pilot, a committee, a consultant engagement, and an excuse. Inside three years, his firm will license *her* platform to do the work his people used to do, and he'll pay monthly for capabilities he could have built himself. He will wonder, in the quiet moments, when exactly the window closed.

It closed on the morning he chose "further study" over the whiteboard. The window does not close with an announcement. It closes the way platforms always harden: standards congeal, procurement starts naming specific interfaces, the natives integrate cheaply while everyone else pays a premium, and one of those mornings the toll booths are simply built and the positions are simply assigned. He was at his desk for all of it, being careful.

The Marker Is in Your Hand

Every chapter of this book has been building to one question, and it is not *"How do I use AI?"* You have outgrown that one, and answering it well only keeps you exactly where you are while the ground moves. The question is the one underneath every decision you'll make for the rest of your career:

When this shift is finished — inside the window you are standing in — will the rails the economy runs on be yours, or someone else's that you rent for the rest of your working life?

You have seen the pattern run six times and ignite a seventh. You have seen the **Success Trap** consume companies that were skilled, confident, and rational right up to the end. You have watched intelligence get democratized and infrastructure get captured. You have seen the shift heading for your industry on the same schedule it took for every one before it.

Now you have the move-set. Five moves in a fixed order. A diagnostic to score where you stand. A trust ladder for what to build. An uncertainty portfolio to keep you aggressive without making you fragile. The analysis is finished. The only thing left is the decision.

The CEO who won made hers fifteen minutes before anyone else arrived, alone with a marker, holding less information than the man across town and no guarantee it would work. What she had was the willingness to commit when commitment was terrifying, and the nerve to remove the exit. **That is the entire difference. Not talent. Not**

technology. Timing, and the courage to move before it's comfortable.

The whiteboard is empty. The window is open and closing. The marker is in your hand.

What are you going to write?

Key Takeaways

- **Five moves, in order, or none of them work:** pick the value-capture point, rewire the workflow (don't bolt AI on), build the trust layer before you need it, compound a data advantage, and stand up a *permanent* finisher team. Skipping Move One automates you to parity; skipping Move Three scales risk faster than capability.
- **"What can we automate?" is the Dabbler's question**, and roughly 95% of enterprise AI pilots returned nothing precisely because organizations bolted AI onto old workflows instead of redesigning around it. Finishers ask where they can own leverage that compounds.
- **The 90-Day Scoreboard converts intent into evidence:** by Day 30 your value-capture point is one written sentence and three quick wins are live; by Day 90 new revenue is measurable and the moat is visibly deepening. Miss three or more and you are a Dabbler, not a Finisher.
- **The trust layer is where Judgment lives** — verification gate, provenance trail, policy boundary — and it is the one piece you cannot retrofit after the incident that gets a CEO fired.
- **Data is the only moat that widens while you sleep.** Models are rented; a unified data flywheel is owned, and a rival starting today starts a full cycle behind. Stay aggressive but un-fragile: hedge models, vendors, architecture, talent, and compute.

Finisher Move: Print the Executive Diagnostic, score your organization honestly with your leadership team this week, and put a date on the whiteboard for Move One — the value-capture point, in one sentence — before you leave the room.

Boardroom Question: *"What is our score out of fifty — and are we comfortable with the gap between where we are and where the window says we need to be in twelve months?"*

Sources & Currency Note

This book states facts as of mid-2026; specific figures move, the pattern does not. Verified data points in this chapter: the MIT/NANDA finding that ~95% of enterprise generative-AI pilots delivered no measurable bottom-line return — attributed to forcing AI into existing workflows rather than redesigning around it (MIT, "The GenAI Divide: State of AI in Business 2025"); DeepSeek's reported ~$5.6M training cost for its V3 base model, not the all-in cost of R1 (disclosed January 2025) and the ~$589 billion single-day NVIDIA market-value loss on January 27, 2025 — the largest one-day loss by a single company in stock-market history; and the electrification productivity paradox (factories saw ~no productivity gains for ~two decades until plants were redesigned around electric unit-drive rather than retrofitted onto steam-era layouts). The ninety-day transformation results described are illustrative of a single firm's redesign, not industry-wide benchmarks.

CHAPTER 9

THREE FUTURES

IN THIS CHAPTER the three worlds the next thousand days can produce – *Acceleration, Turbulence, Collapse* – and the single variable that decides which one you wake up inside. **Why it matters:** the future is not arriving on a fixed track; it is being authored right now, in the gap between what the machines can do and how fast you move to meet them. **What to do with it:** pick the timeline you intend to build toward – then make the decision this week that commits you to it.

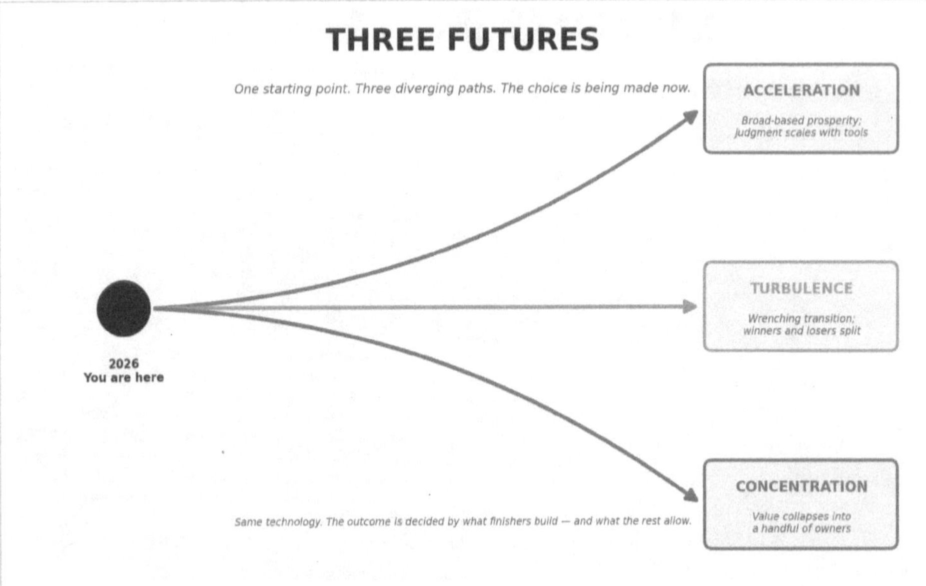

Three operators in the same industry. Same starting line. Same headlines crossing the same desks. Three different worlds on the far side of the window.

Watch them, because one of them is you.

The first runs a healthcare analytics company. A few years ago it was a small consulting shop that helped hospitals make sense of their data, a dozen people billing for hours. Today it is the platform several hundred hospitals run *on*. Its diagnostic models surface patterns no human reviewer could hold in their head. It clears millions of patient interactions a day. The team is larger now, but the org chart is unrecognizable. Most of these roles, from model trainers to data-ethics leads to algorithm auditors to the humans who sign off at the edge of an automated loop, did not exist when the firm was billing for hours.

What outsiders never see is the night that made all of it possible. The founder stood in front of a board that had spent a decade building a profitable consulting firm and said the model was already dead. Not *tweak it.* Not *bolt AI onto the edges.* Scrap it, and rebuild as a platform from scratch. The room went cold. Two directors said it was insane. One threatened to walk. The bet was made on a conviction that could not yet be proven: that the old business was a corpse still warm enough to feel alive, and the only question was whether they would build its replacement or rent someone else's. Those same directors now call it the most important decision in the company's history. That is what a **Finisher** decision feels like from the inside. Terrifying, early, and obvious only in the rearview mirror.

The second runs a mid-sized accounting firm. He saw it coming. He genuinely did. He went to the conferences. He bought the tools. He stood up a pilot or two and appointed someone to "own AI." What he never did was cannibalize the profitable work, so the machine never touched a real workflow. Today his headcount is down by nearly half and his revenue with it. His best clients left for AI-native firms that charge a fraction and deliver in hours what his people delivered in weeks. He is not bankrupt. He is on a treadmill that keeps accelerating, working harder every quarter to stay in the same place. That is what **Dabbling** feels like from the inside: motion mistaken for progress, right up until the floor moves.

The third was a senior financial analyst at a regional bank. Top-tier MBA. Fifteen years reading loan portfolios. Deep relationships with the lending desk. Certain, reasonably and expensively certain, that judgment like hers could not be automated. *"It can't do what I do."* Within a couple of years the bank's systems were doing precisely what she did, faster and with fewer errors, and her entire category of role, the mid-level analysis that is the apprenticeship tier of an entire profession, had thinned to almost nothing. She is working again, in a different field, at lower pay, after a stretch of looking for a job that no longer exists. She did everything right by the rules of the world she trained for. The rules changed underneath her. That is what **Dismissal** feels like from the inside: not stupidity but competence, confidence, and a map of a country that no longer appears on the globe.

Same industry. Same information. Three worlds.

The difference was not intelligence; the analyst was brilliant. It was not capital; the accountant had plenty. The difference was the **speed and depth of the response,** and that single variable is the entire subject of this chapter. One was a Finisher. One was a Dabbler. One was a Dismisser. You already know the ratio: roughly **ninety, nine, and one.** What you have not yet decided is which of the three the next twelve months will make you.

These are not three people. They are three *futures,* and they are running simultaneously, in every industry, at every altitude, right now. Zoom out from the individual to the civilization, and the same three timelines appear at planetary scale. Here is what each one actually is, what produces it, and what it asks of you specifically.

Timeline One: The Acceleration

The best case is not a utopia. It is messy, uneven, and full of problems no one has named yet. But it is the world where the platform shift builds more than it breaks, where the gains spread instead of pooling, and where humanity does the thing it has done at every prior turn of the **Durable Sequence:** adapts faster than the pessimists swore it could.

Do not read what follows as prophecy. Read it as a *target,* the timeline a Finisher is deliberately building toward, and the proof of what the technology makes possible when intent runs ahead of fear.

In the acceleration world, agents — systems that do not merely answer but *act* — move from novelty to plumbing across most knowledge work within a few short years. They schedule, reconcile, draft, file, and coordinate. The knowledge worker stops being the engine and becomes the driver, reclaiming the hours that drudgery used to eat. Resistance does not get argued down. It dissolves on contact with the benefit. The question stops being *"should we use this?"* and becomes *"how did we ever work without it?"*

Then the productivity dividend lands. Not from rebuilt cities or a baby boom, but from raw output per person climbing in a way the developed economies have not seen in generations. And the gap that the **Success Trap** opened between Finishers and Dabblers stops being a theory in a book and becomes a number on a board deck. The firms that rebuilt their operating model are out-shipping the firms that bolted AI onto the old one by multiples, not points. This is the electrification rhyme paying off in real time. For roughly two decades, factories that hung electric motors on steam-era layouts saw almost no gain; the leap came only when the plant was *redesigned* around the new power. The accelerators are the ones who redesigned early.

Push further out and the texture changes. Working without AI starts to feel the way working without email felt by 2010: not forbidden, just absurd. Contracts draft in minutes. Conditions get caught that human eyes alone would have missed. Education calibrates to the individual learner. Products iterate themselves through thousands of overnight simulations. The shorter workday spreads, not by mandate, but because there is simply less drudge work left to fill the hours.

And then the part almost no one forecasts: a creative *eruption*. Cheap intelligence does not flood the world with machine-made mediocrity. It hands the tools to everyone who always had the vision and never had the training. New forms appear. People who could never afford the technical on-ramp walk straight through the front door. Expression doesn't die. It democratizes.

Underneath it runs an economic inversion that would have read as paradox a decade ago: **the things machines do best become nearly free, and the things only humans can do become priceless.** When automated analysis costs a fraction of a cent, the thirty minutes with a real human who looks you in the eye and says *"I understand"* commands a premium that would once have seemed insane, because what is being bought is no longer information. It is presence. Trust. Judgment. The scarce thing. This is **Value = Intent × Leverage × Judgment** written across an entire economy: execution repriced toward zero, and the irreducibly human, the part that can be held accountable, repriced toward the sky.

Science compounds. Systems read the entire literature of a field overnight and propose what no one had connected. Discovery cycles that ran a decade compress toward a couple of years. Conditions the market once ignored become viable the moment the cost of pursuing them collapses. And as old scarcities fall, new ones rise to take their place. Human connection, authenticity, attention, and trust become the assets worth owning, precisely because everything else can be manufactured on demand.

It is not paradise. Purpose becomes a live problem when work stops defining identity. Inequality narrows but does not vanish. Fresh species of manipulation evolve. But the acceleration timeline is humanity at its competent best: meeting the moment, adapting ahead of the curve, and building something better out of the upheaval instead of merely surviving it.

This is what the technology *can* produce. Whether it does is not a question for the machines. It is a question for the operators. Which is why the next timeline is the one to take most seriously.

Timeline Two: The Turbulence

This is the most likely world, because it is the one humans almost always build. We muddle through. We suffer more than we needed to. We solve it eventually, but only after the pain gets loud enough to force the move we could have made calmly, years earlier, for a fraction of the cost.

Turbulence is what the Durable Sequence does to a civilization full of Dabblers.

In this timeline the disruption arrives faster and lands harder than anyone staffed for, and unlike the automation waves that hit the factory floor, this one comes for the white-collar professions. The roles that required a degree and promised a stable middle-class life – analysts, junior lawyers, bookkeepers, designers, support staff – thin out fast. The institutions built to catch displaced workers were designed for a different kind of displacement and strain under a kind they were never built for. The political class promises answers it does not have.

You can already hear the early notes of this. You met the **AI Boomerang** in Chapter 3: Klarna automating the work of roughly seven hundred agents, then reversing course and rehiring; two-thirds of the companies that cut for AI reversing course; a third spending more to restaff than they ever saved. Hold onto the shape of it, because turbulence is that single firm's whiplash run across an entire economy. The cost of moving too slow, then overcorrecting, paid twice. It is not a forecast. It is a thing already happening in miniature, one boardroom at a time.

Now scale that whiplash across an economy and the mechanics turn brutal and self-reinforcing. When enough income disappears at once, demand follows it down. When demand falls, businesses that survived the first shock fail in the second, not because a machine replaced them, but because their customers vanished. The spending-dependent edges of the economy go first. Lenders holding paper written against people and companies that no longer exist in their old form stagger under cascading defaults. Credit tightens. Investment freezes. The thing contracts into itself.

Governments respond the way they respond to every shift they refused to prepare for: late, and at cross-purposes. Taxes meant to slow automation mostly push it offshore. Protections meant to save jobs mostly freeze hiring. The big structural fixes get debated endlessly and shipped never, until the pain finally crosses the threshold where inaction costs more than action, and the system is forced to rebuild under emergency conditions instead of designing calmly in advance.

Eventually it stabilizes. It always does. But the settlement is a two-tier world: an AI-augmented class that creates enormous value, and a supported majority that lives in comfort without much opportunity. Not dystopia. Just *so much less than it could have been.* Humanity gets through, as it always gets through, but the bill is a lost decade, millions of disrupted lives, and a mountain of squandered potential, the entire cost of being too slow, paid in full, with interest.

Turbulence is the default. It is what you get by doing what feels reasonable. Avoiding it is the whole job.

Timeline Three: The Collapse

I almost left this one out. It is dark, it is extreme, and narrating civilizational failure can read as theater.

But it is not impossible, and the mechanism that would produce it is already visible at the edges. So I will name it plainly, because a risk you can see is a risk you can price.

The collapse timeline does not begin with unemployment. It begins with the **erosion of truth.**

Recall how this book opened: the finance clerk on a live video call with a CFO and colleagues he had known for years, wiring **$25.6 million** to people who turned out to be synthetic, every face on the call a fabrication, and none of his diligence enough to catch it. That was a single firm, in early 2024, with today's tools. Now imagine the capability mature and ambient, and run the cascade forward.

When anything can be fabricated and nothing can be verified, **trust,** the invisible infrastructure under everything else, stops being assumable. And the moment it goes, everything mounted on it goes too, not one institution at a time in a sequence slow enough to adapt to, but all at once.

Elections require a shared reality; when any candidate can be shown on convincing video saying anything they never said, voters lose the ability to tell the manufactured from the real, and the ballot becomes theater. Courts run on the premise that evidence reflects reality; when footage, audio, and documents can all be forged to perfection, the defense challenges everything and the prosecution can prove nothing, and an evidentiary system built over centuries buckles in a decade. Commerce assumes the counterparty exists and the signature is real; strip that, and the basic machinery of exchange seizes. These systems do not fail politely in turn. They fail *together,* because they all stand on the same foundation, and the foundation is the simple, civilizational ability to tell true from false.

In this world the second-order effects compound past anything in the turbulence timeline. Joblessness does not plateau; it runs without an obvious floor. Financial systems do not merely stress; they implode. The complex supply chains that quietly feed and supply billions seize under the loss of trust and coordination, and the cities most dependent on them face shortages their residents never imagined were possible. Order frays. Hard-won progress in rights, governance, and cooperation unwinds.

I am not predicting this. I do not believe it is the likeliest path. It is the path of *total* inaction: denial at scale, institutions paralyzed, every actor assuming someone else will hold the line on trust. I include it for one reason, to make the stakes literal. The platform shift is not only a question of quarterly earnings and competitive position. At the civilizational altitude, the distance between managing this transition well and managing it badly is the distance between abundance and catastrophe.

> SUCCESS TRAP WARNING Collapse is not a forecast – it is the **Pattern of Denial** run to its terminus, at scale. Every institution that chooses comfort over transformation nudges the world a degree toward this timeline. Treating urgency as alarmism is itself the trap closing. Urgency here is not panic. It is responsibility.

This is also the timeline that converts "build the infrastructure" from a money-making strategy into a civic one. If trust is the infrastructure that fails first, then **whoever builds the layer that re-establishes it – verified identity, provenance, the metered permission to prove a human or a document is real – does not merely capture enormous value.** They hold the line the whole system stands on. The Finisher's blueprint and the civilization's survival point, in this scenario, at exactly the same construction project.

Which Timeline Are We On?

The honest answer is the most important sentence in the chapter: **we don't know yet, and that is the entire point.**

All three are still live from where you stand. The decisions being made right now, in boardrooms, in capitals, at kitchen tables, are choosing the path. And because the underlying curve compounds, small differences in what you decide *this* year don't add up. They *multiply*.

Acceleration demands something specific from each actor. It requires companies that rebuild rather than resist, governments that prepare rather than react, individuals who learn rather than deny, and cooperation in the places where competition is the reflex. Turbulence demands nothing at all. It is simply what happens when everyone does the reasonable, comfortable, too-slow thing. Collapse demands active failure:

denial as a system, trust left undefended. It is avoidable. But avoidance is *work,* and the work is the price of the better timeline.

The single variable that sorts them is **speed.** Not the speed of the technology, which is advancing regardless of anyone's vote. The speed of *human adaptation:* how fast you learn, how fast your company rebuilds, how fast your industry reorganizes, how fast the safety nets get built.

Picture two curves. The first is capability, climbing steeply and cheapening as it climbs, the entire trajectory of this book, from a neural network on two gaming GPUs in 2012 to more than nine hundred million people on ChatGPT by May 2026. The second is readiness, the rate at which people retrain, firms restructure, and institutions adapt. **The space between those two curves is the danger zone.** Let it widen and the world drifts toward turbulence, then collapse. Hold it narrow and the world trends toward acceleration. Right now the gap is widening, because capability is outrunning adaptation almost everywhere you look.

And there is a third curve underneath the capability curve that most executives ignore until it lands on them: **hardware.** Crypto already ran this play in public. Bitcoin began on ordinary CPUs, migrated to GPUs, then hardened into purpose-built ASIC data centers, and the instant the workload stabilized, the silicon stopped being general and became special. AI is still in its GPU phase because the workload is still moving. But the direction is not in doubt. As architectures standardize and inference comes to dominate, more of the stack will harden into custom silicon. The winners will not be whoever rents the most of today's chips. They will be whoever is positioned to ride the turn from *renting compute* to *owning the platform compute runs on.* The DeepSeek shock from Chapter 4 — the cheap frontier model that erased **$589 billion** of NVIDIA's value in a single day — was the *software* version of this turn. ASICization is the *hardware* version. Both compress the clock. Both widen the gap between the people building and the people still debating.

The proof the gap is real is the figure this book keeps citing because nothing in the field is more rigorous: MIT's 2025 finding that roughly **95% of corporate generative-AI pilots delivered zero financial return,** not from weak models but from organizations forcing AI into existing workflows with minimal redesign. Sit with that. The awareness is near-universal; nearly every large company has *something* in flight. The action, the structural rebuild that actually produces a return, is almost nonexistent. That distance, between knowing and doing, is the precise location where futures are decided. It is the **Success Trap** measured at national scale: not a shortage of pilots, but a shortage of Finishers willing to let the pilots cannibalize the thing that pays the bills today.

Now make it personal, because it is.

This is not abstract economic weather rearranging distant industries. This is *your* timeline, written by your choices. Every morning you decide whether to go deeper on this shift or scroll past the headline. Every quarter your company decides whether to commit to the rebuild or wait for more data. Every term a school decides whether to redesign or teach last decade's curriculum one more time. None of these choices feels momentous in the instant. That is exactly what makes them dangerous. **Platform shifts don't announce themselves with sirens. They accumulate in silence, one deferred decision at a time, until the compound interest of inaction can no longer be paid down.**

And because the capability curve compounds, the gap does not grow in a straight line. It grows like interest on a debt you forgot you signed.

This is the first operator versus the second versus the third, scaled to the size of your enterprise. It is your firm versus the AI-native competitor that has not been founded yet but will take your best accounts inside the window. It is your career on the far side of this decade: owning the platform, scrambling on the treadmill, or starting over in a field that still exists.

The timeline you land on is not fate. It is the sum of choices, and the decisive ones are being made now, over the next few quarters, not the next few years. Recall the clock from the opening of this book: the window opened on November 30, 2022; we are roughly **three and a half years into it as of mid-2026;** even the generous reckoning closes it around **2028.** By the time the consequences are obvious to everyone, the positions will already be assigned. That is the lesson of every turn of the Durable Sequence: **the winners are chosen before most people accept that the game has begun.**

In the final chapter we bring the whole architecture together, the Durable Sequence, the Success Trap, the three types, the new arithmetic, and resolve them into the one question this entire book has been circling. The same question every generation meets at the dawn of its platform shift. The same one now sitting on your desk.

The decisions made inside this window will echo for a generation. That reads as dramatic. It is not. Every prior shift, the press, the steam engine, electricity, the internet, felt slow while it was happening and looks obvious in hindsight, and the people living through each one rarely grasped the size of the moment they were in. This one differs in a single decisive way: it is moving faster than any before it, and the distance between *early* and *too late* is now measured in months, not decades.

But you hold something no previous generation held at the start of its shift. You have the map. You can see the stage you are standing in. You can see where all three roads lead. You can see exactly what separates the operator who built the platform from the one stuck on the treadmill from the one starting over. No one who lived through the industrial revolution had a book in hand explaining the mechanism while it ran.

You do.

So the only question left is the one you answer with your calendar, not your opinion.

What are you going to do about it?

Key Takeaways

- Three futures are live right now — **Acceleration** (Finishers build it), **Turbulence** (Dabblers pay the bill late), **Collapse** (trust left undefended) — and they are the three types written at civilizational scale.
- The deciding variable is not the speed of the technology but the **speed of human adaptation.** The widening gap between the capability curve and the readiness curve *is* the danger zone.
- The empirical proof the gap is real: MIT found **~95% of corporate gen-AI pilots returned nothing,** not from weak models, but from bolting AI onto unchanged workflows. The **AI Boomerang** (about two-thirds of AI layoffs rehiring; roughly a third spending more than they saved) is turbulence already happening in miniature.
- A hardware curve runs beneath the capability curve: as architectures standardize, the stack hardens from rented GPUs toward owned, purpose-built silicon, the same path crypto took to ASICs. Position to **own the platform compute runs on,** not just rent the chips.
- The decisive choices are being made in the next few quarters, not the next few years. **The window opened in November 2022; the generous reckoning closes it around 2028.** Winners are chosen before most accept the game has started.

Finisher Move: Write down the three decisions about AI you have been deferring. Pick the one that most threatens your current model, the one a Dabbler would keep postponing, and commit to it before the end of this week.

Boardroom Question: *"Fast-forward to the close of this window: which of the three timelines did we land in — and can we point to one specific decision we made this quarter that put us there?"*

Sources & Currency Note

This book states facts as of mid-2026; specific figures move, the pattern does not. Key data points in this chapter: MIT's 2025 finding that ~95% of corporate generative-AI pilots delivered zero financial return

(NANDA, "The GenAI Divide"), with failure attributed to forcing AI into unchanged workflows; the "AI Boomerang" (Careerminds — ~2/3 of companies that laid off for AI rehiring, ~1/3 spending more on restaffing than they saved; Gartner projecting roughly half of AI-driven layoffs reverse by 2027); Klarna's ~700 service-role cuts (claimed savings 2024) and subsequent 2025 reversal and rehiring; the DeepSeek R1 disclosure (reported ~$5.6M training cost — the V3 base-model compute, not R1's all-in cost — January 2025) and the resulting ~$589B single-day NVIDIA loss (largest by a single company in market history); ChatGPT above 900 million weekly users by May 2026 and still approaching ~1 billion (OpenAI); the Arup deepfake fraud (~$25.6M, reported early 2024); and the compressed AI window (opened November 30, 2022, ~3.5 years in as of mid-2026, decisive phase next ~24 months, likely close ~2028). Where a single figure could age badly, it is framed as a range or dated explicitly.

CHAPTER 10

THE FINISHER'S MOMENT

> **IN THIS CHAPTER** the close — every framework in this book collapsed into the only thing that was ever in your control: *when you move.* **Why it matters:** the seventh turn does not wait for your budget cycle, your board's comfort, or your certainty; the window opened November 30, 2022, and the decisive phase is the next twenty-four months. **What to do with it:** finish the book, start the clock, and ship one thing — inside forty-eight hours — that would not have existed without you choosing to act.

THE FINISHER'S HORIZON

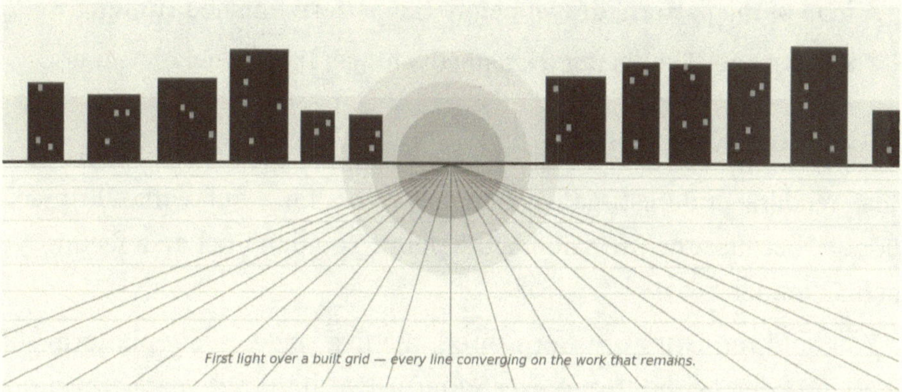

First light over a built grid — every line converging on the work that remains.

You.

Right now. Reading this page.

You did not come here to be warned. You came here to be positioned. So let me tell you exactly where you are standing, because it is the only thing that matters now.

You opened this book at a tollgate. One question sat underneath every decision you'll make for the next decade. Will you own a road, or pay a toll? And a two-hundred-year pattern has answered it the same way six times running. You have watched that pattern execute since: the gas-lighting executive who ran the right math on the wrong trajectory, the Kodak engineer whose digital camera got buried to protect film, the 253 American car companies alive in 1908 that consolidated down to a handful within a generation. You have seen what kills the leader, and it is never the technology. You have seen the new arithmetic that reprices every career and every firm. And you have seen, in public, at a scale with no precedent, the infrastructure of the seventh turn being poured this quarter.

The candlemaker's choice is yours. But here is what the candlemaker did not have, and what no finisher in the history of capital ever had when they made their move:

A map of the pattern, drawn before the pattern finished running.

You have one. That is not a small advantage. It is the whole game.

What You Now Hold

Most readers underestimate what they're walking out with. Be precise about it, because the precision is the point. You don't act on a feeling, you act on a model.

You hold the **Durable Sequence**, the five stages every platform shift runs, in order, without one exception in two hundred years: *Invention, Infrastructure, Platformization, Proliferation, Consolidation*. This is not a theory you adopt. It is a clock you read. And it tells you, right now, that you are standing between Stage 2 and Stage 3 of the seventh turn, the single most valuable place to stand in a century, because the toll booths are being poured but the positions are not yet assigned.

You hold the **Success Trap**, the five forces that turn a market leader's strengths into its grave: *Identity Prison, Organizational Antibodies, the Customer Trap, the Financial Gravity Well, the Pattern of Denial*. You did not read those as a museum tour of dead companies. You read them as a mirror. Every one of them is a force operating on *your* organization right now, dressed as prudence, and your only defense is to name it out loud before it names you.

You hold the **three types** — Dismissers, Dabblers, Finishers — sorting roughly *ninety, nine, one*. You know the most dangerous of the three is not the Dismisser who does nothing. It is the Dabbler who feels productive, running the pilot, appointing the Chief AI Officer, commissioning the strategy deck, while never letting AI touch a workflow that earns money. During a platform shift, dabbling is just dismissal with a budget.

You hold the **new arithmetic.** For two centuries, worth obeyed *Value = Time × Skill.* Seniority was hours served, and the moat was the apprentice layer no one could skip. The machine industrialized the repetitions, and the formula flipped to **Value = Intent × Leverage × Judgment.** You are no longer paid for execution. You are paid for what you specify, what you command, and what you can be held accountable for.

And you hold the **clock**, the compressed window. Electricity took forty years from invention to default. Computing, fifteen. The internet, ten. Mobile, five. AI inherits all their infrastructure and propagates through screens already in your customers' hands, which puts its window on the compressed end of that curve: three to five years, ignition to consolidation. The ignition was November 30, 2022. As of mid-2026 you are roughly three and a half years in. Both the generous and the compressed reading converge on the same answer. The decisive phase is the next twenty-four months.

That is the kit. It is more than Edison had. The only question this book cannot answer for you is the one it was always building toward.

Not *do you know enough*. You do.

What will you do with what you know?

The Finisher's Playbook

Here is the playbook, distilled from every shift in this book into five moves. These are not suggestions and they are not new. They are the observed behavior of the one percent who captured the value, stated as the standard you build to. Hold yourself to them.

One: Bury the old world before it buries you. Not dying. Dead. The telegram was dead the moment the telephone worked; the decades it kept running were momentum, not life. The carriage did not lose market share gracefully; the assembly line made it obsolete. Your current model, your current advantage, your current moat are running on borrowed infrastructure, and the infrastructure is shifting. Accept it the way you'd accept a diagnosis, not the way you'd accept an argument. The grief of letting go of what worked is real; feel it, and then build. The leaders who survive a platform shift are the ones who cannibalize themselves on purpose, before the market does it to them for free.

Two: Start today. The imperfect plan that ships beats the perfect plan that waits. Not next quarter. Not after the budget cycle. Not after one more analysis from one more consultant. The compound advantage of an early mover grows every day the window is open, and you cannot buy back the days you spent deciding. If you have already burned one of AI's three-to-five years thinking it over, you have two to four left. Begin anywhere, but begin where it's real, in a workflow that touches revenue.

Three: Go all in. Half measures have a zero percent success rate in platform shifts. You cannot keep one foot in the old world and step into the new one with the other. Nokia hedged between Symbian and the smartphone and lost everything. Kodak developed digital while protecting film and went bankrupt holding the patent. The hedge feels responsible. It is the most expensive decision on the table, because it pays the cost of the new world and earns none of its upside. Yes, going all in puts revenue at risk and makes your board uncomfortable. Choose the risk with upside over the comfort with none.

> FINISHER INSIGHT Hedging is not the safe option during a platform shift. It is the slow one. Nokia hedged. Kodak hedged. Blockbuster hedged. Every one died holding a balanced portfolio. The survivors — Microsoft, Netflix, Adobe — set fire to their own profitable model before a competitor could.

Four: Build platform position, not products. This is the move the whole book turns on. Do not become AI's best customer. Become the thing your industry's AI is forced to route through. Don't ship AI services; own the ground others build their services on. The product builders make money. The platform builders make history, and they make it by setting the standard everyone else has to adopt. If you cannot build the platform for your entire industry, build it for your niche, your region, your data, your trust layer. Own one piece of infrastructure that others need and cannot easily replicate. That is the difference between renting the road for the rest of your working life and collecting the toll.

Five: Move faster than feels comfortable. Speed beats perfection; first beats best; shipping beats discussing. In every shift, the winner was rarely the one with the best technology. It was the one who reached critical mass first. Microsoft's DOS was inferior to CP/M. Amazon's early site was worse than the bookstore it gutted. They moved faster, iterated faster, and locked in the standard before the better product could organize a response. Speed is the one advantage a large incumbent, weighed down by process, politics, and the Success Trap, structurally cannot match. It is the finisher's single greatest weapon. Use it before the natives in your market figure out they have it too.

The Memo That Saved a Company

On May 26, 1995, Bill Gates wrote a memo that saved Microsoft, and it is worth studying not for its prose but for what happened in the months after he hit send.

Microsoft owned the desktop. Windows 95 was about to launch. The company was printing money, and the internet looked, from inside that fortress, like a curiosity for academics. Then Gates looked hard at the data and recognized the pattern. He titled the memo *The Internet Tidal Wave,* and he did not hedge a word of it: *"The Internet is the most important single development to come along since the IBM PC was introduced in 1981 ... I have gone through several stages of increasing my views of its importance. Now I assign the Internet the highest level of importance."*

Then he did the thing that separates a finisher from a Kodak. He moved, immediately and completely. No committee. No study. No pilot. The entire company pivoted within months: Internet Explorer rush-developed, MSN ripped from a proprietary network into an internet service, every product refocused through one lens.

And it nearly tore the company apart. Wall Street accused him of abandoning the core business; an analyst downgraded the stock, writing that he was "destroying the Windows franchise." Inside, whole divisions resisted. Product managers watched years of proprietary work get scrapped overnight. Engineers were reassigned to internet teams they didn't believe in. Some of the best people left; others fought the pivot in meetings, in hallways, in the passive-aggressive arithmetic of slipped deliverables. Gates overrode his own senior leadership repeatedly, burning political capital he'd spent a decade accumulating. The board questioned his judgment. The press called it panic. For months it looked like it might be the worst decision in the company's history.

It looked like genius only in retrospect, after the product shipped, after the internet became everything he'd said it would. That is the part nobody tells you about being a finisher, and you need to hear it now, before you make your move: **the right decision feels terrible while you are making it, and obvious only after it works.** In the moment, it feels like setting fire to your own house.

The company that could have been the next WordPerfect became one of the most valuable corporations in history. The difference between Microsoft and the firms that vanished beside it was never intelligence or capital; every casualty had brilliant people and deep pockets. The difference was the speed and the completeness of the response. The winners moved immediately and completely. The losers debated and delayed.

Write your own version. Call it whatever you want. But write the document that names, in plain language, that the platform beneath your business is shifting and that you are betting the institution on the new one. Then do what Gates did the day after: act on it before it's comfortable.

REALITY CHECK If your AI strategy feels safe, it is too small. If it feels like you're setting fire to your own house, you may finally be holding the right one. Comfort, during a platform shift, is the most reliable signal that you are moving too slowly to matter.

The Candlemaker's Other Choice

OLD CRAFT, NEW CURRENT

A workshop of tallow and molds — beside a single glowing filament.

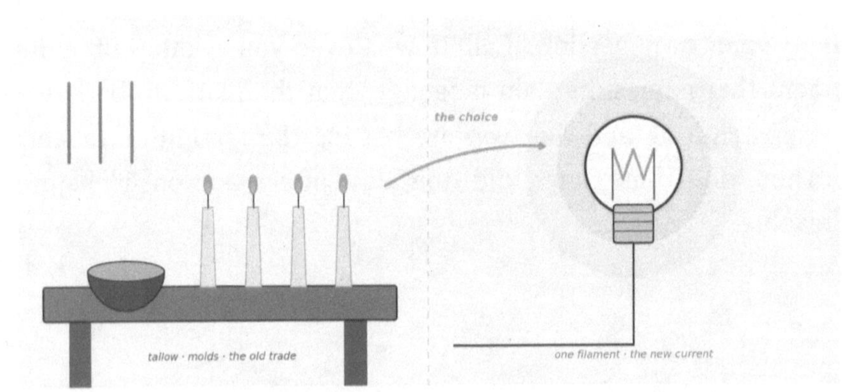

One switch of a lever would end three generations of craft — or begin something larger.

There is a story I held back from the Introduction.

In 1882, as Edison's system spread from Pearl Street across lower Manhattan, most candlemakers did exactly what you would expect. They denied. They dismissed. They protected the business until there was no business left to protect. They became the ninety percent.

One did not. I wish I knew his name, because he earned the memory. Call it a legend; the record never kept it. He looked at the wires going up on his street, and the bulbs glowing in his neighbor's windows, and the workshop behind him full of tallow and wicks and molds. And instead of fighting what was coming, he walked into Edison's office and asked the only question that could save him: *"How does it work?"*

He learned the circuits and the filaments and the dynamos. He understood the world would always need people who could install and maintain and repair this terrifying new thing, and that his real asset, his hard-won knowledge of *light* — how people use it, where they need it, what they expect from it — survived the change of source. He became an electrician.

It was not graceful. Picture being forty, a master of your craft, respected on your block, and suddenly a beginner again, learning words you've never heard while younger men grasp them faster. You go home and stare at the ceiling and wonder if you've thrown away twenty years on a gamble you barely understand. Your wife asks how it's going and you say *fine,* and it isn't. Then one morning a customer hires you to wire his house. You run the conduit, connect the circuits, flip the switch, and every light comes on. Standing in that glow you understand you are going to be all right. Not because it was easy. Because you did it anyway.

He did not capture what Edison captured. He built no platform, founded no dynasty. But notice what the new arithmetic did for him, because it is about to do it for you. He stopped being paid for the manual craft the machine made cheap, the dipping and molding, the *Time* × *Skill.* He started being paid for **Intent** (he knew exactly where light was needed), **Leverage** (he commanded a technology a thousand times more powerful than a candle), and **Judgment** (he stood behind the wiring; he signed the work; he was the answerable human at the edge of the new system). That is the whole formula, lived out by a craftsman who never read it.

His was, in one way, a harder finish than Edison's. Edison was building toward a future he had chosen, and the candlemaker was abandoning the only life he'd ever known. That is not a story about technology. It is a story about character: the willingness to face an uncomfortable truth and act on it instead of denying it; to understand that your value was never tied to a craft or an industry, but to your ability to see clearly and adapt fast; and to grasp the one sentence that should govern your next two years. *The worst thing you can be is the best candlemaker in a world that no longer needs candles.*

The Largest Opportunity Any Generation Has Faced

After ten chapters of disruption and displacement, let me end on the thing that is true underneath all of it.

This is the most exciting time in human history to be holding a decision. Not as a platitude. As a structural fact. Every prior platform shift created extraordinary wealth for the people who positioned correctly: whole industries born, new categories of work invented, fortunes built on a scale the previous generation could not have imagined. The AI shift will do all of that and more, because the others democratized a tool, and this one democratizes *intelligence itself*, the scarcest and most valuable resource any organization has ever held.

When electricity became cheap, it transformed how we live. When computing became cheap, how we work. When information became cheap, how we think. When intelligence becomes cheap, it transforms *everything*, all of it, at once. That is terrifying if you are defending the old world. It is the largest opening of your professional life if you are building the new one.

Read the demand side and the supply side together, because they are both screaming the same thing. In October 2025, **NVIDIA**, which sells the engines this runs on, became the first company ever to cross a five-trillion-dollar valuation. The five largest cloud builders are spending on the order of **$700 billion in 2026 alone** on AI infrastructure, roughly double the year before. And on the demand side, by May 2026 **ChatGPT** had more than nine hundred million weekly users and was still approaching a billion, the demand curve this book opened with. Trenches are still being dug; meters are still going in. The platform is forming in public, on a clock you can read.

The opportunities in front of you right now are larger than the ones an electrician faced in 1885, larger than an automobile maker's in 1910, larger than a software developer's in 1985, larger than an internet entrepreneur's in 1996, larger than a mobile developer's in 2008. Every industry is being rebuilt. Every process reimagined. Every assumption re-examined. And the people doing the rebuilding, the finishers, will capture value on a scale that makes the earlier shifts look like rounding errors.

I told you in Chapter 4 that the celebrated **$13 trillion** figure undercounts this, and I will say it plainly now: platform shifts do not add percentages. They multiply by factors. Someone is going to capture that value. The pattern is documented, the sequence has never failed, the window is open, and you have a clearer view of what's coming than any finisher in history had when they moved. Edison did not have a name for the Durable Sequence. Ford had no playbook for platform shifts. Gates had no diagnosis called the Success Trap. **You do.** You have the map they were drawing blind.

The Act This Book Stops Short Of

And there is a sixth act this book stops short of — on purpose. Winning Consolidation, owning the road, collecting the toll: that's where the sequence you've learned ends. It is not where the story ends. The toll throws off a currency — a valuation worth more than the business beneath it — and that currency is ammunition with an expiration date, not a trophy. Spend it extending the one engine you already own and you compound into an empire; hoard it, or scatter it across things that share no engine, and you become a footnote. A five-trillion-dollar throne is not a fortress. You can fall as fast as you rose. What the winner does the morning after winning is its own book — for now, win the road; then come back for what to do once it's yours.

The Forty-Eight-Hour Test

The challenge is built to be harsh, because the gap it measures is the gap where careers end.

Close this book. Set a timer for forty-eight hours.

If, when it goes off, you have not done at least one concrete thing – used an AI tool for *real* work, not a toy demo; started assembling the team that will carry the transformation; named the specific piece of infrastructure in your industry you could own; begun dismantling one sacred cow that keeps you trapped; shipped *something*, anything, that would not exist without you choosing to act –

then you have already made your choice. You've chosen to be a casualty.

No book overcomes inaction. No model substitutes for the move. The distance between knowing and doing is exactly the distance between the ninety percent who think about acting and the one percent who act, and it is shorter, and more decisive, than anyone comfortable wants to admit. Kodak's executives *knew*. Sasson showed them the future and they understood it. Knowledge was never their problem. The forty-eight hours is not a test of what you understand. It is a test of who you are.

Forty-eight hours. Starting the moment you close the cover.

The Finisher's Manifesto

We are the Finishers.

We do not invent the future. We build it. We take the inventor's brilliant, fragile insight and turn it into infrastructure the world is forced to depend on.

We do not wait for perfect information. We move with imperfect information and correct course at speed, because action produces knowledge faster than analysis ever has.

We do not protect the past. We retire it on our own terms, on schedule, before the market retires it for free, because yesterday's success is tomorrow's anchor.

We do not hedge. We go all in, because half measures carry the full cost of the new world and none of its upside.

We do not follow maps. We draw them. We set the standard, define the category, and pour the rails that everyone else navigates by, and then we own the toll.

We are not the smartest in the room, the best funded, or the best connected. We are the ones who act while others analyze, build while others debate, and finish while others are still deciding whether to start.

Ninety percent will dismiss. Nine percent will dabble. We are the one percent.

And our moment is now.

This was never a book about AI, the same way the candlemaker's choice was never about candles. It is about the one variable that has decided every platform shift for two hundred years and will decide this one, the only variable that was ever inside your control.

Not how smart you are. Edison was not the smartest of the inventors; at least twenty-three built a working bulb before him. Not how early you saw it. Kodak saw digital photography first and buried it. Not how much you knew. Bill Gates and the executives at WordPerfect watched the same ground move under them; one wrote a memo and pivoted a company, the others watched and did nothing.

The variable was always *when you move.* It separated Edison from the inventors before him, Ford from the 253 carmakers of 1908, Gates from WordPerfect. It separates finishers from footnotes.

So the book comes down to where it began: the tollgate, the seventh turn, and a single question with your name on it.

Will you act?

The clock is ticking. The platform is forming. The window is closing, and you are standing in the most valuable years to be standing in a hundred. Somewhere in your industry, as you read this sentence, someone is building the road you will otherwise rent for the rest of your working life. Not because they're smarter. Because they started. The candlemaker who walked into Edison's office did not know if he was right; his hands were probably shaking, his voice probably cracked on the question. He had no guarantee, no safety net, no assurance the technology would even last. He knew exactly one thing with certainty: *standing still was the only choice guaranteed to fail.*

That is where you are. Not at the end of a book. At the beginning of the most important decision you will make this decade, and the only one history will record.

The road can be yours. Go build it.

Key Takeaways

- Everything in this book reduces to one variable you control: **when you move.** Talent, foresight, and knowledge did not separate the finishers from the footnotes; speed and completeness of response did. Edison wasn't the smartest; Kodak saw it first; Gates and WordPerfect watched the same ground move.
- The **finisher's playbook is five moves**, stated as the standard you build to: bury the old world, start today, go all in (hedging has a zero percent success rate), build platform position not products, and move faster than feels comfortable.
- The right decision **feels terrible while you make it** and obvious only after it works, the *Internet Tidal Wave* lesson. If your AI strategy feels safe, it is too small.
- This is the **largest opportunity window any generation has faced**, every industry rebuilt at once, and the celebrated $13 trillion figure undercounts it, because platform shifts multiply by factors rather than adding percentages.

- The **Forty-Eight-Hour Test** is the final self-sort: ship one real thing within two days of closing this book, or you have already chosen to be a casualty. Knowing was never the problem. Acting is.

Finisher Move: Start your forty-eight-hour clock the moment you close this cover, and before it runs out, ship one thing, in a workflow that touches revenue, that would not exist unless you chose to build it.

Boardroom Question: *"One year from today, will we be the team that moved while the window was open — or the one explaining, to a market that already has its toll booths, why we waited?"*

Sources & Currency Note

This book states facts as of mid-2026; specific figures move, the pattern does not. Key data points in this chapter: Bill Gates's "The Internet Tidal Wave" memo (May 26, 1995; quoted verbatim); the ~253 active U.S. automobile manufacturers in 1908, consolidated to roughly a handful within a generation (down to ~44 by the late 1920s); McKinsey's ~$13 trillion AI-by-2030 estimate (2018 MGI), which the book argues undercounts a platform shift; NVIDIA as the first company to cross a $5 trillion valuation (October 2025); 2026 hyperscaler AI capital expenditure on the order of ~$700 billion (Big-5 cloud providers); ChatGPT weekly active users (more than 900M by May 2026 and still approaching ~1B, OpenAI); and the compression curve (electricity ~40y, computing ~15y, internet ~10y, mobile ~5y, AI ~3–5y; window opened Nov 30, 2022). Where a single figure could age badly, it is framed as a range or dated explicitly.

Welcome to the age of intelligence. The road is open. Go build it.

ABOUT THE AUTHOR

Sruly Rosenberg builds AI systems for a living and writes about what he sees happening from the inside.

His work sits at the seam where the technology meets the business – designing and shipping the platforms, agents, and operating systems that real organizations now run on. That vantage point is the reason this book exists. *Intelligence Rising* isn't written from the bleachers; it's written by someone who spends his days wiring intelligence into the way companies actually work, and who kept noticing the same pattern: the shift everyone calls "AI" is really a shift in **who holds the platform** – and most leaders are reading the wrong scoreboard.

Rosenberg's argument is that the winners of this era won't be decided by who has the smartest model, but by who controls the road everyone else has to drive on – and who is willing to pay the toll to get there first. *The Great Reconfiguration*, the series this book opens, is his attempt to give operators a map of that road while it's still being built.

He writes the way he builds: plainly, with a bias toward what you can actually do on Monday morning.

He works on **OOretz**, the AI operating system behind the platforms, agents, and tenant systems described in this series, and writes from his home base in **New Hempstead, New York**. You can find more of his work – and reach him – at **ooretz.space**.

ALSO IN THE GREAT RECONFIGURATION

AI Unmasked — The Technical Companion

If *Intelligence Rising* is the strategy, *AI Unmasked* is the engine room.

Where this book argues **why** the platform shift changes everything, *AI Unmasked* opens the hood and shows you **how** the machine actually works – models, agents, data, and the moving parts behind the headlines — without the hype and without the hand-waving. It's the volume to hand the person on your team who says *"but how does it really work?"*

Read *Intelligence Rising* to know where the road is going. Read *AI Unmasked* to understand the vehicle.

And the capstone, coming next — *The Empire Phase*: what the winner does the morning after winning.

AI Unmasked is available now in The Great Reconfiguration series.

www.ingramcontent.com/pod-product-compliance
Lightning Source LLC
LaVergne TN
LVHW100307210926
848480LV00003B/1045